# Environmental Footprints and Eco-design of Products and Processes

**Series Editor**

Subramanian Senthilkannan Muthu, Head of Sustainability - SgT Group and API, Hong Kong, Kowloon, Hong Kong

This series aims to broadly cover all the aspects related to environmental assessment of products, development of environmental and ecological indicators and eco-design of various products and processes. Below are the areas fall under the aims and scope of this series, but not limited to: Environmental Life Cycle Assessment; Social Life Cycle Assessment; Organizational and Product Carbon Footprints; Ecological, Energy and Water Footprints; Life cycle costing; Environmental and sustainable indicators; Environmental impact assessment methods and tools; Eco-design (sustainable design) aspects and tools; Biodegradation studies; Recycling; Solid waste management; Environmental and social audits; Green Purchasing and tools; Product environmental footprints; Environmental management standards and regulations; Eco-labels; Green Claims and green washing; Assessment of sustainability aspects.

More information about this series at http://www.springer.com/series/13340

Subramanian Senthilkannan Muthu
Editor

# Carbon Footprints

Case Studies from the Building, Household, and Agricultural Sectors

 Springer

*Editor*
Subramanian Senthilkannan Muthu
Head of Sustainability - SgT Group and API
Hong Kong, Kowloon, Hong Kong

ISSN 2345-7651          ISSN 2345-766X   (electronic)
Environmental Footprints and Eco-design of Products and Processes
ISBN 978-981-13-7918-5          ISBN 978-981-13-7916-1   (eBook)
https://doi.org/10.1007/978-981-13-7916-1

This Springer imprint is published by the registered company Springer Nature Singapore Pte Ltd.
The registered company address is: 152 Beach Road, #21-01/04 Gateway East, Singapore 189721, Singapore

# Contents

# The Carbon Footprints of Agricultural Products in Canada

R. L. Desjardins, D. E. Worth, J. A. Dyer, X. P. C. Vergé
and B. G. McConkey

**Abstract** The rapid increase in the atmospheric concentrations of greenhouse gases (GHGs) has given rise to international commitments to reduce GHG emissions such as the Kyoto Protocol and the Paris Climate Agreement. If countries are going to be successful in meeting their commitments and help reduce the impact of climate change, it is essential that all sectors of the economy be part of the solution. In Canada, the agriculture sector, which encompasses a wide array of production systems and commodities, accounts for about 12% of the anthropogenic GHG emissions. Numerous techniques have been developed to quantify GHG emissions from the agriculture sector. The data collected have been used to calculate the carbon footprints of a wide range of agricultural products and to develop indicators to help formulate climate change mitigation and adaptation policies for the sector. It is well known that agricultural soils have lost large quantities of carbon in the past. With some of the changes in management practices, agricultural soils in some regions have now become significant sinks of carbon. The agriculture sector is responsible for carbon dioxide emissions associated with fertilizer production, farm fieldwork operations, machinery supply and a variety of other smaller sources. It is also the biggest anthropogenic source of methane and nitrous oxide. If we are to promote the consumption of low carbon products, it is then important to have an accurate estimate of the GHG emissions associated with their production. Carbon footprint estimates vary substantially depending on the units and what is included in the calculations. Recent estimates of the carbon footprints per unit protein for animal products ranged from 215 kg $CO_2e$ for sheep to 15 kg $CO_2e$ for poultry–broiler meat. As expected, the carbon footprints per unit protein of plant products are substantially less. Some examples are presented on how the sharing of the environmental burden reduces the magnitude

R. L. Desjardins (✉) · D. E. Worth · X. P. C. Vergé
Ottawa Research and Development Centre, Agriculture and Agri-Food Canada,
960 Carling Avenue, Ottawa, ON K1A 0C6, Canada
e-mail: ray.desjardins@canada.ca

J. A. Dyer
Agro-Environmental Consultant, Cambridge, ON, Canada

B. G. McConkey
Swift Current Research and Development Centre,
Agriculture and Agri-Food Canada, Swift Current, SK, Canada

© Crown 2020

S. S. Muthu (ed.), *Carbon Footprints*, Environmental Footprints and Eco-design
of Products and Processes, https://doi.org/10.1007/978-981-13-7916-1_1

of the carbon footprints of certain products and how environmental indicators can be used to develop policies. These results highlight opportunities for climate change mitigation by consumers and producers of agricultural products in Canada.

**Keywords** Emissions intensity · Carbon footprint · Agricultural products · Protein · Soil carbon · Greenhouse gas

## 1 Introduction

Due to the rapid increase in the atmospheric greenhouse gas (GHG) concentrations, most sectors of the economy must find ways to reduce their emissions (IPCC 2013). There has been an increase in global temperature of almost 1 °C since 1900 (Hansen et al. 2010). The international community has begun to mobilize resources to meet international agreements on reducing GHG emissions to attempt to minimize the impact of climate change. As part of the United Nations Framework Convention on Climate Change, the Paris Agreement was recently signed by more than 190 nations. This agreement, which is the first global treaty to try to combat climate change, is an attempt to slow down the increase in the atmospheric concentration of GHGs and to hold the global increase in air temperature to below 1.5 °C. If this target is to be achieved, it is generally accepted that GHG emission reductions greater than current Nationally Determined Contributions as a part of the Paris Agreement are required (Rogelj et al. 2016; van Soest et al. 2017). All sectors of the economy will need to contribute by improving their efficiency and reducing their GHG emissions. This fact has led many researchers to calculate the carbon footprints associated with a wide range of products and projects; several examples are presented in this book. The carbon footprints of agricultural products are a function of the management practices, location and types of production systems. There is real interest in using the carbon footprints to inform food choices to reduce GHG emissions through diet (Clark and Tilman 2017; Pradhan et al. 2013).

In Canada, the agri-food sector's anthropogenic GHG emissions are reported annually by Environment and Climate Change Canada (Environment and Climate Change Canada 2017). Agriculture is a relatively unique sector, because not only does it contribute to climate change, it is also very sensitive to it. The sector consists of a wide array of production systems and commodities supported by a vast and diverse set of landscapes. Canada's heavy reliance on livestock production makes it challenging for carbon footprint calculations (Vergé et al. 2012). The need for accurate estimates of the carbon footprints of agricultural products becomes even more important when the use of agricultural products for bioenergy production is considered. The great challenge for the agriculture sector, in the coming decades, will be to meet the increasing global food demand and yet minimize GHG emissions. According to Tilman et al. (2011), a strategic intensification of agriculture that elevates yields from existing croplands can meet most of the 2050 global food demand and lead to considerably less GHG emissions than an approach requiring land clearing.

We will describe the measuring and modeling techniques used to quantify GHG emissions from the agriculture sector. We will present $CO_2$, $CH_4$ and $N_2O$ emission estimates for the main agricultural sources in Canada. Agricultural soils, which in the past have lost large amounts of carbon, can now be part of the solution because they have become significant sinks of carbon in some regions. We will present carbon footprint estimates for a wide range of Canadian agricultural products. We will show how these estimates vary spatially and temporally and that good knowledge of the carbon footprints of agricultural products could help policy makers and consumers make recommendations and food and fuel choices that would lead to a more sustainable agricultural sector. We will present examples of how the sharing of the environmental burden influences the magnitude of the carbon footprint of certain products. We will also discuss how environmental indicators could be used to develop policies to mitigate climate change.

## 2 Overview

### 2.1 Canadian Agriculture

Agriculture covers about 5% of Canada's land mass and produces a variety of crops and livestock. Total farmland, in the 2016 Census of Agriculture in Canada, was reported to consist of approximately 70 Mha of which 35 Mha was intensively cultivated (Statistics Canada 2018a). A series of maps presented by Janzen et al. (1999) show the percentage of farmland as a proportion of total land area, the percentage of pasture (improved and unimproved) as a proportion of farmland, the percentage of annual crops as a proportion of farmland and the distribution of livestock on an animal unit basis. The Prairies, which account for about 80% of Canada's farmland, consist mainly of croplands and grasslands with sections of concentrated livestock production. Canola, wheat, barley, oats, flax, lentils, peas and sunflowers are the main crops. Corn and soybeans are mainly grown in Ontario and Québec. Potatoes are mainly grown in the Maritime provinces. Tree fruits, small fruits and wine are mainly produced in British Columbia, the Niagara Peninsula of Ontario and southwestern Québec. Vegetables are grown all across the country particularly near big cities. Canning crops such as peas, tomatoes and corn are produced mainly in Southern Ontario and Southern Alberta. Irrigation in Canada occurs predominantly in southern Alberta, with lesser amounts in Saskatchewan and in the valleys of British Columbia. It is also used for high-value crops such as market garden crops and small fruits. Animal husbandry is practiced in all provinces. In 2016, there were 12.5 million head of cattle, of which 0.9 million were dairy cows (Statistics Canada 2018b), 14 million hogs (Statistics Canada 2018d), 145 million poultry (Statistics Canada 2018f), 1 million sheep and lambs (Statistics Canada 2018c) and 0.3 million horses and ponies (Statistics Canada 2018e).

## 2.2 Main Agricultural Sources and Sinks of GHGs

Agriculture emits the three main anthropogenic GHGs ($CO_2$, $N_2O$ and $CH_4$). The main sources and sinks of these gases are shown in Fig. 1. Agricultural lands can absorb, as well as emit, all three gases. The relative magnitude of the fluxes is shown by the length of the arrows. Fluxes of $N_2O$ and $CH_4$ are usually expressed as $CO_2$ equivalent ($CO_2e$) in order to account for the relative impact by which these three gases trap heat. On a time horizon of 100 years, a molecule of $N_2O$ has 298 times more impact on radiative forcing than $CO_2$, while a molecule of $CH_4$ has 25 times more impact than $CO_2$ (IPCC 2007).

## 2.3 Measurement Techniques

Canadian scientists have developed a whole series of techniques to measure agricultural sources and sinks of GHGs that cover a wide range of spatial and temporal scales (Fig. 2). Soil cores have been collected for many years to quantify the amount and change of soil organic carbon (SOC) in agricultural soils (e.g., Campbell et al. 2000). Because SOC is highly variable spatially, it is necessary to have careful measurements involving a large number of soil samples to obtain an accurate estimate of

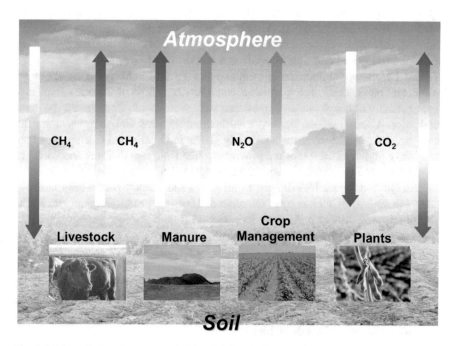

**Fig. 1** Main agricultural sources and sinks of GHGs in Canada

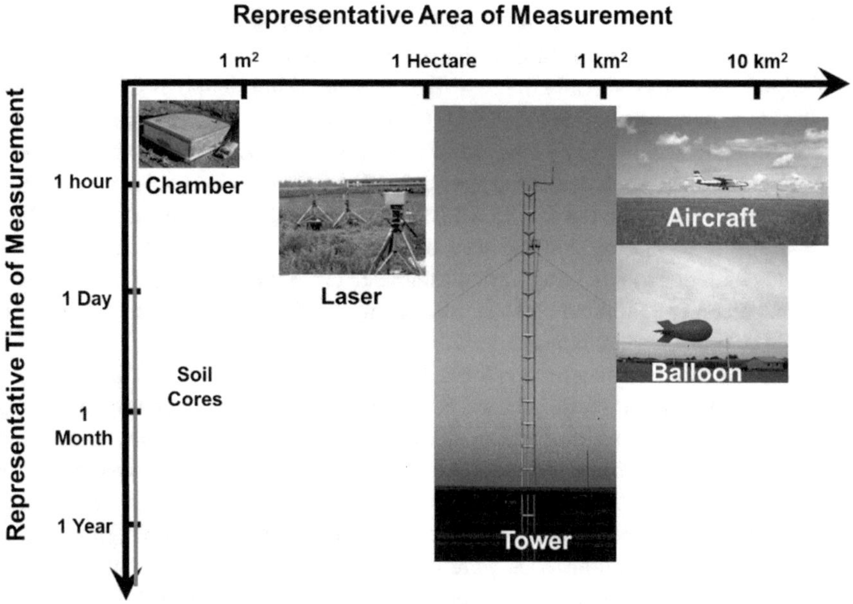

**Fig. 2** Spatial and temporal scales of techniques used to measure GHG emissions from agricultural sources

SOC change. With a representative average, it is then possible to estimate the change in C over time and/or between different land uses and managements. Chamber techniques have been widely used to quantify emissions and uptake of $CO_2$, $CH_4$ and $N_2O$ from and to agricultural soils. There are basically two types of chambers: (1) steady-state chambers where the air flows through the chamber at a constant rate. In this case, the concentration of the gas in the chamber is constant (2) non-flow through non-steady-state (NFT-NSS) chambers where the flux density is estimated by calculating the rate of change of the concentration of the gas of interest. In a study based on NFT-NSS chamber measurements of soil $N_2O$ flux density, Rochette and Eriksen-Hamel (2008) reported that while the measurements obtained were valid for comparison between treatments within a study, they were often biased estimates of the actual emissions. Micrometeorological techniques such as the mass balance and inverse modeling technique have been used to quantify GHG emissions from local sources. Desjardins et al. (2004) demonstrated that both techniques gave similar results but because of the versatility of the inverse modeling technique it is more frequently used for on-farm measurements. VanderZaag et al. (2014) used this technique to obtain continuous measurements of $CH_4$ emissions from barns and manure storage systems. They reported 40% higher $CH_4$ emissions in the fall when the liquid manure storage was full. During that period, 60% of the farm emissions came from the manure storage and the rest from the animals in the barn. Tower-based flux systems, which provide continuous flux measurements at the field scale, have been used to quantify mass and energy exchange over many different ecosystems

(Baldocchi et al. 2001). Pattey et al. (2006a); (Pattey et al. 2007) used tower-based flux systems to quantify the $N_2O$ emissions associated with spring thaw; these are quite significant but practically impossible to measure using chambers without affecting the measurements. Aircraft-based $N_2O$ flux measurements in combination with model estimates were used by Desjardins et al. (2010a) to obtain regional estimates of $N_2O$ emissions. They estimated that the indirect $N_2O$ emissions for the region that they studied accounted for 23% of the agricultural $N_2O$ emissions, which is close to the indirect emission estimate of 20% suggested in the IPCC methodology. The challenge of verifying agricultural $CH_4$ emission inventories at a regional scale was recently examined by Desjardins et al. (2018). They showed that aircraft-based $CH_4$ flux measurements agreed reasonably well with the agricultural $CH_4$ emission inventory estimates when the $CH_4$ sources within the aircraft flux footprint were mainly from agricultural sources. However, in many cases, the comparison was complicated by the presence of unaccounted for $CH_4$ sources such as wetlands, waste treatment plants and farm-based biodigestion of animal waste. Measuring the change in the concentration of $N_2O$ in the nocturnal boundary layer profile over time has been shown to provide a reasonable estimate of the magnitude of $N_2O$ (Pattey et al. 2006b) emissions. The emissions were estimated by calculating the change in the $N_2O$ concentration in the nocturnal boundary layer two hours apart. As demonstrated by Desjardins et al. (2019), the inversion layer at night acts as a cover and restricts gas from escaping.

## 2.4   Modeling GHG Emissions

Due to the complexity of agricultural production systems, the translation of field measurements of GHG emissions into spatial and temporal estimates of GHG emission intensities requires an assortment of models. Figure 3 shows the range of complexity and intended users for some of the models used in Canada. Process-based models, which are highly complex, have been used to generate farm, regional and national estimates of soil C change (Smith et al. 1997, 2001), $N_2O$ (Grant et al. 2004; Smith et al. 2004) and $CH_4$ emissions (Ellis et al. 2009; Guest et al. 2017; Li et al. 2012). They have been used to determine Canada-specific emission factors that have been incorporated in the Intergovernmental Panel on Climate Change (IPCC) Tier II methodology to calculate soil C change as well as $N_2O$ and $CH_4$ emissions at provincial and national scales. GHG emission estimates combined with production estimates have been used to calculate the carbon footprints of the major crops in Canada (Dyer et al. 2010b). The Unified Livestock Industry and Crop Emission Estimating System (ULICEES), which uses the carbon footprint of crops and animal production estimates, has been used to estimate the carbon footprints of Canadian livestock products (Vergé et al. 2007, 2008, 2009a, b, 2012, 2013). It has also been used to assess trade-off for Canadian agriculture such as reallocating the protein production from a ruminant to a non-ruminant source (Dyer and Vergé 2015). Vergé et al. (2012) recognized the difficulty that soil carbon is a sink term, whereas the other

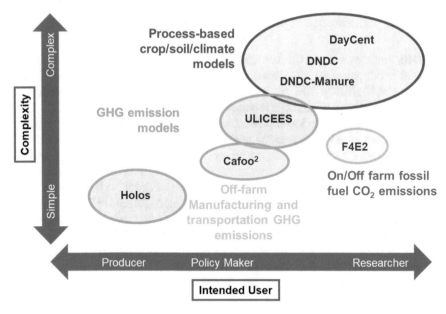

**Fig. 3** Models used to estimate the agricultural GHG emissions and the carbon footprints of agricultural products in Canada

terms in the GHG emissions budget were annual fluxes. Hence, annual $CO_2$ fluxes to and from the soil vary depending on the state of the soil carbon sink (discussed in more detail in Sect. 3.2). This difficulty was dealt with in ULICEES (Vergé et al. 2012) by defining the payback periods for annual GHG emissions to account for quantities of sequestered soil carbon.

Two other models have been used to calculate the carbon footprint of agricultural products in Canada. The Canadian Food Carbon Footprint (Cafoo$^2$) calculator has been used to estimate off-farm GHG emissions associated with the production of the main dairy products (Vergé et al. 2013). On- and off-farm fossil fuel $CO_2$ emissions have been estimated using the F4E2 model (Dyer and Desjardins 2003, 2005b). This model, which includes 22 energy terms in the farm field operations, has been used to estimate the direct and indirect energy-based GHG emissions from Canadian agriculture (Dyer et al. 2014b). Several sub-models have recently been added to F4E2 to provide a complete all-commodity picture of the Canadian farm energy and fossil $CO_2$ emission budget (Dyer and Desjardins 2009, 2018; Dyer et al. 2011b). Holos, which is a farm model, is being used to demonstrate to farmers how their on-farm GHG emissions are affected by the decisions that they make (Janzen et al. 2006; Kröbel et al. 2013). All these models have different levels of complexity and intended users.

## 3  GHG Emissions from the Agriculture Sector in Canada

In 2016, Environment and Climate Change Canada (ECCC) estimated the total agricultural anthropogenic GHG emissions at 60 Mt $CO_2e$ (Environment and Climate Change Canada 2017): 50% were in the form of nitrous oxide ($N_2O$) and 50% in the form of methane ($CH_4$). The reported GHG emission estimates by ECCC from the agriculture sector are incomplete. Carbon dioxide emissions from agricultural fossil fuel use are attributed to the energy and transportation sectors, while $CO_2$ emissions from agricultural lands are reported in the land use, land-use change and forestry (LULUCF) sector under the cropland category. In order to account for the full impact of the agriculture sector, these GHG emissions must be included when estimating the carbon footprints of agricultural products. As seen in Fig. 4, including the 3% of the emissions attributable to fossil fuel, the total emissions from the agriculture sector account for 12% of Canada's GHG emissions. As more measurements of GHG emissions from the agriculture sector become available, emission estimates are improving but because of the large variety of crops and animal products, the wide range of management practices and the variability of soil and climatic conditions the estimates are highly uncertain (Karimi-Zindashty et al. 2012).

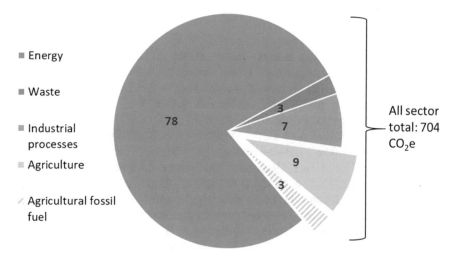

**Fig. 4**  Percentage contribution of sectors of the economy to Canada's GHG emissions in 2015

## 3.1  Magnitude of the GHG Emissions from the Agriculture Sector

The relative magnitude of the GHG emissions in 2015 from the agriculture sector in Canada is presented for the three main GHGs (Fig. 5). Carbon dioxide emissions from fossil fuel use accounted for 26% of the emissions. Nitrous oxide emissions accounted for 36%, while $CH_4$ accounted for 38%. The main sources of $CO_2$ from fossil fuel use are for producing fertilizer (32%) followed by field operations (25%), machinery supply (17%), farm transport (7%), electricity generation (6%) and heating (5%) (Dyer and Desjardins 2005a; Worth et al. 2016). As production efficiencies improve, the emissions associated with some of these products are decreasing. Based on a fertilizer manufacturing analysis by Nagy (2000), $CO_2$ emissions associated with the supply of farm chemical inputs were estimated as $4.05$ t $CO_2$ $t^{-1}$ N (Dyer and Desjardins 2009; Dyer et al. 2014a). Snyder et al. (2007) reported $CO_2$ to nitrogen conversion rates that were the same as the $4.05$ t $CO_2$ $t^{-1}$ N conversion reported by Dyer et al. (2014b) for Nebraska and 10% higher for Michigan. This makes 4.05 a mid-range GHG emission cost based on these sources. Canadian fertilizer industrial sources suggest an N fertilizer supply coefficient closer to 3.1 for Canada. Whereas part of this difference could be attributed to an increase in the efficiency of producing ammonia, this lower coefficient only includes the manufacture of N fertilizer. The $4.05$ t $CO_2$ $t^{-1}$ N reported by Dyer et al. (2014b) includes phosphate and potash fertilizers and the supply of pesticides, as well as the transport of all farm chemicals to the farm. In view of the importance of this term to the farm energy budget, the

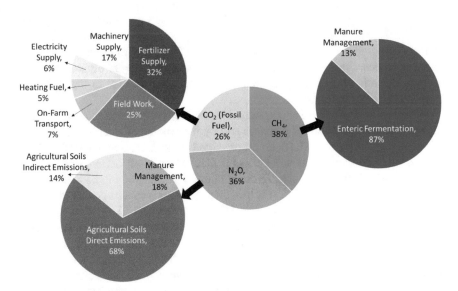

**Fig. 5** Relative magnitude of the sources of $CO_2$, $N_2O$ and $CH_4$ emissions from the agriculture sector in Canada in 2015

efficacy of this industrial claim should be evaluated by an objective third party using peer-reviewed scientific methodology. Nitrous oxide emissions are mainly associated with agricultural soils. About 68% are direct emissions, and 14% are indirect emissions. Manure management accounts for the remaining 18%. Methane emissions are dominated by enteric fermentation which accounts for 87% of the emissions, while manure management accounts for the remainder.

## 3.2   Soil Carbon in Agricultural Soils

Agricultural soils can either be a source or a sink of carbon. Land clearance for agriculture is estimated to have been a large source of $CO_2$. According to Ruddiman (2003) between 8000 years and 200 years ago, the small annual $CO_2$ emissions prior to the Industrial Revolution contributed about two times more $CO_2$ than the post-industrial revolution emissions. Soils are now recognized for their potential to sequester carbon. Cole et al. (1997) estimated that globally 0.4–0.8 Pg C year$^{-1}$ could be sequestered for 50–100 years. Lal and Bruce (1999) estimated that the world croplands have the potential to sequester the equivalent of 50% of the annual emissions by deforestation and other agricultural activities. Desjardins et al. (2005) and Desjardins et al. (2001) discussed and presented the amount of C sequestered in agricultural soils in Canada for a whole series of management practices. They reported that the main opportunities for increasing soil C sequestration were converting croplands to grasslands, reduced tillage, reduction of bare fallow and introducing forage in crop rotations. They ran the Century and DNDC models for five locations across Canada for a 30-year time period. As expected, the models predicted that conversion of croplands to grasslands would result in the largest reduction in GHG emissions. Soil carbon values associated with the change in: (1) mixture of cropland types, (2) tillage practices, (3) area of summer fallowing, (4) cultivation of organic soils, (5) perennial woody crops and (6) residual emissions are presented in Fig. 6 for the 1981–2015 period in Canada.

Including SOC change in footprints adds challenges because the rate of SOC change is affected by current practices and by the amount of total SOC, the latter being a legacy of past LUM. If current practice is associated with increasing SOC, e.g., no-till, then those gains will be less if SOC was high. If current practices are associated with losing SOC, e.g., frequent fallow, those losses will be greater if the SOC was high. However, if SOC was low, then the gains would be greater and losses smaller. Since there is a wide range in SOC from the past LUM, then there is a wide range in gains or losses for any specific current land use and management. Consequently, unlike other GHG emissions for other resources used in production for which representative GHG emissions can be practically estimated, it is generally impractical to estimate the soil C change in any one year for all land parcels involved in production in that year. Hence, a more generic approach is most commonly used for agricultural products derived from a large area. The SOC change across a whole agricultural area that is involved in production is estimated from generic change such

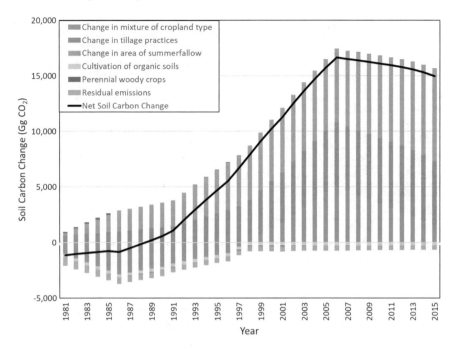

**Fig. 6** Soil carbon change in agricultural soils in Canada, 1981–2015, due to changes in management practices (Environment and Climate Change Canada 2017)

as regional emission factors (Goglio et al. 2015; Lindorfer et al. 2014; Shrestha et al. 2014). This can be considered the soil C implications for land occupation (Müller-Wenk and Brandão 2010; Schmidinger and Stehfest 2012; Wiedemann et al. 2015), and so the same C change value can be attributed for all products derived from that land similarly occupied by agriculture. There is more uncertainty regarding including land-use change (LUC). The time frame for considering past LUC is uncertain but can affect the C change greatly (Goglio et al. 2015; Hörtenhuber et al. 2014). The indirect land-use change that is imputed from global land competition can vary greatly depending on methodology and assumptions (Flysjö et al. 2012; Mathews and Tan 2009). Normal practice is to report the soil C implications separately so the user can decide how to treat it for the intensity estimates.

Table 1 illustrates the estimated losses and gains of SOC in 2015 within Canada for changes between annual crops and perennial forage/pasture, for changes in tillage intensity and for changes in the area of fallow as estimated for Canada's national GHG emission inventory (Environment and Climate Change Canada 2017). In all provinces, there are important C changes from changes between annual crops and perennial forages/pastures, but these area changes occur in both directions so there are both increases and decreases in SOC. In Eastern Canada, conversion from perennial forage/pasture to annual crops dominates and that causes agricultural land in general to lose SOC. Changes in tillage and fallow are principally toward reductions and so

**Table 1** Average soil C change in 2015 for agricultural land within Canada

| Province | Soil organic carbon change[a] (kg C ha$^{-1}$) due to land management change | | | | | | |
|---|---|---|---|---|---|---|---|
| | Annual to perennial | Perennial to annual | Reduce tillage intensity | Increase tillage intensity | Decrease fallow | Increase fallow | Net |
| Atlantic provinces | 78.7 | −141.5 | 2.4 | −0.2 | 7.4 | −4.4 | −57.7 |
| Québec | 45.2 | −187.9 | 6.8 | −0.1 | 5.3 | −2.9 | −133.5 |
| Ontario | 36.7 | −169.2 | 10.6 | −1.4 | 6.0 | −1.8 | −119.1 |
| Manitoba | 50.9 | −43.4 | 22.1 | −2.6 | 42.2 | −8.1 | 61.0 |
| Saskatchewan | 59.6 | −28.6 | 37.7 | −0.4 | 74.0 | −10.9 | 131.3 |
| Alberta | 57.1 | −52.3 | 27.8 | −0.1 | 38.3 | −6.1 | 64.6 |
| British Columbia | 85.5 | −75.1 | 5.4 | −0.2 | 25.3 | −12.7 | 28.2 |

[a]Positive is an increase, negative is a decrease, and multiply by 3.67 to convert to $CO_2$

represent a SOC increase. These reductions are larger in Western Canada so the SOC on the agricultural land in these provinces is increasing.

## 3.3 The Carbon Footprints of Agricultural Products

The carbon footprint of an agricultural product is based on the calculation of the GHG emissions resulting from the production of the product. The actual carbon footprint value depends on the units used and what is included in the estimate. In most cases, all the GHG emissions to get the product to the farm gate are included. Many carbon footprint values have been published for specific agricultural products. For example, Hillier et al. (2009) used farm survey data from east of Scotland combined with published estimates of $CO_2$ emissions for individual farm operations to determine the carbon footprints of crops such as legumes, winter and spring wheat, oilseed rape and potatoes. They reported that about 75% of the total emissions resulted from nitrogen fertilizer use. The following subsections provide examples of GHG emission intensities for a wide range of Canadian agricultural products. How these different GHG intensity estimates fit together to provide an effective GHG emission policy framework will be discussed in Sect. 3.4. The carbon footprint values can be reduced through the use of renewable resources and more efficient production practices.

### 3.3.1 Carbon Footprints of the Main Agricultural Crops in Canada

The magnitude of the carbon footprints of agricultural products is defined as the total amount of GHG emissions and removals per unit of production. Liang et al.

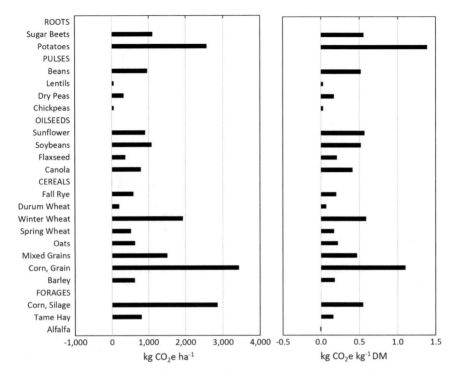

**Fig. 7** GHG emissions kg $CO_2$e per unit area and per kg of dry matter for 21 important crops in Canada in 2011

(2016) emphasized the importance of effective crop rotation systems for increasing crop production, improving soil carbon storage and reducing the carbon footprints of crops. Figure 7 shows the average estimates of the GHG emissions on an area and mass basis for two root crops, four pulse crops, four oilseed crops, eight cereal crops and three forage crops for the year 2011. Data from Statistics Canada were used for production and fertilizer types. In this case, the changes in soil carbon have been included. The emission intensities on an area basis ranged from $-10$ to $3430$ kg $CO_2$e ha$^{-1}$. Negative values occur for crops requiring low inputs, in regions with high soil carbon sequestration, while the heavily fertilized crops such as corn and potatoes have higher carbon footprints in contrast to crops such a soybean and alfalfa that can fix nitrogen and have smaller carbon footprints. When implementing a suite of improved farming practices, negative carbon footprint estimates have been reported for spring wheat, a relatively high-input crop, in the semiarid portions of the Canadian prairies (Gan et al. 2014). On a dry matter (DM) basis, the range is smaller ($-0.01$ to $1.39$ kg $CO_2$e kg$^{-1}$ DM) because a heavily fertilized crop such as corn does not have the highest carbon footprints, as it is a very productive crop.

Tables 2 and 3 present the carbon footprints for the 21 crops shown in Fig. 7. In this case, the values are presented by provinces. These estimates also include soil

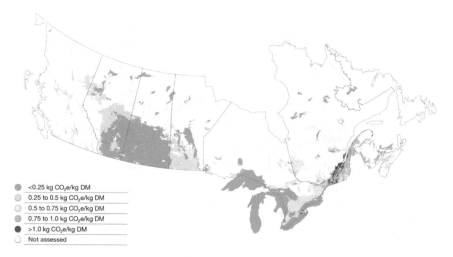

**Fig. 8** Carbon footprints (kg $CO_2$e $kg^{-1}$ DM) for barley crops at the SLC scale, 2011 (Desjardins et al. 2016)

carbon change. We again see a larger range for crops on an area basis than on a DM basis. Large differences are observed at the provincial scale. For instance, the emission intensities in 2011 for canola on an area basis were 2700 kg $CO_2$e $ha^{-1}$ for Québec and 530 for Saskatchewan, while on a DM basis they were 1.23 kg $CO_2$e $kg^{-1}$ DM for Québec and 0.29 for Saskatchewan. For the crops grown nationwide, such as barley, the yields are generally higher in Eastern Canada than in the Prairie Provinces. However, their GHG emission intensities are also higher because the wetter climate in Eastern Canada tends to have higher $N_2O$ emissions, and soils in Eastern Canada tend to be losing carbon primarily as a result of the conversion of perennial to annual cropping.

An example of the actual spatial variations of carbon footprints for the barley crop is presented in Fig. 8. It shows the range in terms of kg $CO_2$e $kg^{-1}$ DM at the SLC scale for all of Canada for 2011. Several factors contribute to the spatial differences. The Prairie Provinces have experienced high adoption rates of practices such as no-till and reduced summer fallowing which favor the sequestration of carbon. These practices, combined with a climate that leads to lower $N_2O$ emissions, and large field sizes that permit more efficient use of farm machinery and reduced fossil fuel consumption, have reduced the agricultural carbon footprints. Therefore, in Western Canada emission intensities are often less than 0.25 kg $CO_2$e $kg^{-1}$ DM. In Eastern Canada, the wetter climate tends to increase $N_2O$ emissions and the smaller field sizes lead to less efficient use of fossil fuel. Additionally, in Eastern Canada there has been a net loss of soil carbon, primarily associated with an increase in the area of annual crops, such as corn and soybean, at the expense of perennial forage crops. These factors tend to cause emission intensities above 0.50 kg $CO_2$e $kg^{-1}$ DM.

**Table 2** Carbon footprints (kg $CO_2e$ $ha^{-1}$) by the province for 21 Canadian field crops in 2011

| Group averages | Oilseeds | Pulses | Roots | Forages | Cereals | Oilseeds | Canola | Flaxseed | Soybeans | Sunflower |
|---|---|---|---|---|---|---|---|---|---|---|
| Atlantic provinces | 860 | | 3300 | 1440 | 2030 | | | | 860 | |
| Québec | 1970 | | 3530 | 1780 | 2680 | | 2700 | | 1240 | |
| Ontario | 1980 | 1360 | 2890 | 1770 | 2330 | | 2740 | | 1220 | |
| Manitoba | 680 | 350 | 1500 | 650 | 1000 | | 1000 | 650 | 140 | 910 |
| Saskatchewan | 400 | 90 | 700 | −180 | 160 | | 530 | 270 | | |
| Alberta | 900 | 530 | 1340 | 580 | 770 | | 1010 | 780 | | |
| British Columbia | 1580 | | 2380 | 1200 | 1180 | | 1580 | | | |
| Canada | 790 | 360 | 1840 | 1220 | 1180 | | 790 | 370 | 1090 | 910 |

| Pulses and roots | Chickpeas | Dry peas | Beans | Lentils | Potatoes | Sugar beets | Forages | Alfalfa | Tame hay | Corn silage |
|---|---|---|---|---|---|---|---|---|---|---|
| Atlantic Provinces | | | | | 3300 | | | 380 | 1320 | 2610 |
| Québec | | | | | 3530 | | | 580 | 1240 | 3530 |
| Ontario | | | 1360 | | 2890 | | | 600 | 1310 | 3390 |
| Manitoba | | 460 | 230 | | 1500 | | | −120 | 510 | 1550 |
| Saskatchewan | 20 | 180 | | 60 | 700 | | | −380 | 30 | |

(continued)

**Table 2** (continued)

| Pulses and roots | Chickpeas | Dry peas | Beans | Lentils | Potatoes | Sugar beets | Forages | Alfalfa | Tame hay | Corn silage |
|---|---|---|---|---|---|---|---|---|---|---|
| Alberta | 430 | 700 | 470 | | 1560 | 1110 | | −130 | 390 | 1480 |
| British Columbia | | | | | 2380 | | | 140 | 850 | 2600 |
| Canada | 60 | 330 | 970 | 60 | 2570 | 1110 | | −10 | 810 | 2860 |

| Cereals | Barley | Grain corn | Mixed grains | Oats | Spring wheat | Winter wheat | Durum wheat | Fall rye |
|---|---|---|---|---|---|---|---|---|
| Atlantic provinces | 1650 | 2780 | 1650 | 1960 | 2020 | 2130 | | |
| Québec | 2250 | 3630 | 2270 | 2590 | 2610 | 2700 | | |
| Ontario | 1930 | 3480 | 1940 | 2240 | 2100 | 2450 | | 2200 |
| Manitoba | 830 | 1550 | | 900 | 890 | 950 | | 880 |
| Saskatchewan | 180 | | | 200 | 170 | 160 | 140 | 130 |
| Alberta | 690 | 1430 | 750 | 690 | 690 | 640 | 630 | 600 |
| British Columbia | 1140 | | | 1180 | 1210 | | | |
| Canada | 620 | 3430 | 1510 | 630 | 520 | 1930 | 200 | 590 |

**Table 3** Carbon footprints (kg $CO_2$e kg$^{-1}$ DM) by the province for 21 Canadian field crops in 2011

| Group averages | Oilseeds | Pulses | Roots | Forages | Cereals | Oilseeds | Canola | Flaxseed | Soybeans | Sunflower |
|---|---|---|---|---|---|---|---|---|---|---|
| Atlantic provinces | 0.45 | | | | | | | | 0.45 | |
| Québec | 0.90 | 0.74 | 1.92 | 0.34 | 0.95 | | 1.23 | | 0.56 | |
| Ontario | 0.95 | | 1.58 | 0.32 | 0.71 | | 1.31 | | 0.58 | |
| Manitoba | 0.42 | 0.19 | 0.82 | 0.15 | 0.41 | | 0.62 | 0.40 | 0.08 | 0.57 |
| Saskatchewan | 0.22 | 0.05 | 0.39 | −0.05 | 0.06 | | 0.29 | 0.15 | | |
| Alberta | 0.41 | 0.27 | 0.67 | 0.14 | 0.21 | | 0.46 | 0.36 | | |
| British Columbia | 0.99 | | 1.30 | 0.25 | 0.41 | | 0.99 | | | |
| Canada | 0.43 | 0.19 | 0.98 | 0.23 | 0.38 | | 0.41 | 0.21 | 0.52 | 0.57 |

| Pulses and roots | Chickpeas | Dry peas | Beans | Lentils | Potatoes | Sugar beets | Forages | Alfalfa | Tame hay | Corn silage |
|---|---|---|---|---|---|---|---|---|---|---|
| Atlantic provinces | | | | | 1.80 | | | 0.08 | 0.26 | 0.52 |
| Québec | | | | | 1.92 | | | 0.11 | 0.24 | 0.68 |
| Ontario | | | 0.74 | | 1.58 | | | 0.11 | 0.23 | 0.61 |
| Manitoba | | 0.25 | 0.12 | | 0.82 | | | −0.03 | 0.12 | 0.37 |
| Saskatchewan | 0.01 | 0.10 | | 0.03 | 0.39 | | | −0.11 | 0.01 | |

(continued)

**Table 3** (continued)

| Pulses and roots | Chickpeas | Dry peas | Beans | Lentils | Potatoes | Sugar beets | Forages | Alfalfa | Tame hay | Corn silage |
|---|---|---|---|---|---|---|---|---|---|---|
| Alberta | 0.22 | 0.35 | 0.24 | | 0.78 | 0.56 | | −0.03 | 0.10 | 0.36 |
| British Columbia | | | | | 1.30 | | | 0.03 | 0.18 | 0.54 |
| Canada | 0.03 | 0.17 | 0.52 | 0.03 | 1.39 | 0.56 | | −0.01 | 0.16 | 0.55 |

| Cereals | Barley | Grain corn | Mixed grains | Oats | Spring wheat | Winter wheat | Durum wheat | Fall rye |
|---|---|---|---|---|---|---|---|---|
| Atlantic provinces | 0.58 | 0.97 | 0.58 | 0.68 | 0.70 | 0.74 | | |
| Québec | 0.80 | 1.29 | 0.81 | 0.92 | 0.93 | 0.96 | | |
| Ontario | 0.59 | 1.06 | 0.59 | 0.68 | 0.64 | 0.74 | | 0.67 |
| Manitoba | 0.34 | 0.64 | | 0.37 | 0.37 | 0.39 | | 0.36 |
| Saskatchewan | 0.06 | | | 0.07 | 0.06 | 0.05 | 0.05 | 0.04 |
| Alberta | 0.19 | 0.40 | 0.21 | 0.19 | 0.19 | 0.18 | 0.18 | 0.17 |
| British Columbia | 0.39 | | | 0.41 | 0.42 | 0.18 | | |
| Canada | 0.18 | 1.10 | 0.47 | 0.22 | 0.17 | 0.59 | 0.07 | 0.20 |

**Table 4** Average carbon footprints per unit area and per unit weight for fruits and vegetables in Canada, 2007–2016

| Vegetables | AP | QC | ON | BC | AP | QC | ON | BC |
|---|---|---|---|---|---|---|---|---|
| | kg $CO_2$e ha$^{-1}$ | | | | kg $CO_2$e kg$^{-1}$ fresh weight | | | |
| Carrots | 3100 | 3100 | 3400 | 3000 | 0.1 | 0.1 | 0.1 | 0.1 |
| Sweet corn | 2300 | 2600 | 2900 | 2600 | 0.4 | 0.3 | 0.2 | 0.4 |
| Tomatoes | 3700 | 4000 | 7000 | 4200 | 0.3 | 0.2 | 0.1 | 0.2 |
| Peas | 1300 | 1400 | 1500 | 1500 | 0.6 | 0.3 | 0.3 | 0.3 |
| Lettuce | 2500 | 3200 | 3100 | 3500 | 0.4 | 0.1 | 0.2 | 0.1 |
| Cabbage | 3800 | 4400 | 4000 | 3800 | 0.1 | 0.1 | 0.1 | 0.2 |
| Potatoes | 2900 | 2800 | 2600 | 2500 | 0.1 | 0.1 | 0.1 | 0.1 |
| *Fruits* | | | | | | | | |
| Blueberries | 1700 | 1500 | 1400 | 1700 | 1.2 | 1.5 | 0.4 | 0.3 |
| Peaches | 2200 | n/a | 1900 | 2100 | 0.2 | n/a | 0.2 | 0.2 |
| Apples | 1800 | 1600 | 1400 | 1800 | 0.1 | 0.1 | 0.1 | 0.1 |
| Strawberries | 2300 | 2200 | 1900 | 2100 | 0.4 | 0.3 | 0.4 | 0.4 |
| Grapes | 1500 | 1400 | 2100 | 1800 | 0.4 | 0.5 | 0.2 | 0.3 |

n/a—Not applicable, no significant production in this region

### 3.3.2 Carbon Footprints for Fruit and Vegetable Crops

Dyer and Desjardins (2018) recently estimated the $CO_2$ emissions for energy used for fruit and vegetable crops. They separated the fruit crops into five groups: (1) apples, (2) stone fruits, (3) bush berries, (4) thorn berries and (5) grapes. They separated the vegetable crops into six groups: (1) roots and tubers, (2) sweet corn, (3) fruit tissue, (4) pulses, (5) leaves and stems, and (6) heads. They treated potatoes separately. The same breakdown will be used to present the carbon footprints of these fruit and vegetable crops. The other main sources of GHG are the $N_2O$ emissions. As in Dyer and Desjardins (2018), it will be assumed that these crops are irrigated, with the exception of potatoes, so the $N_2O$ emissions will mainly be a function of the fertilizer applied and the crop residues produced. There are few estimates of the carbon footprints of fruits and vegetables in the literature, one exception being Stossel et al. (2012), who presented estimates for a large Swiss retailer. Their estimates are slightly larger than the values presented for Canada. The difference can largely be explained by a relatively high $N_2O$ emission factor and the fact that Stossel et al. (2012) included greenhouse production, whereas the results in Table 4 are exclusively for field grown products. Röös et al. (2010) reported that carbon footprint estimates for table potatoes grown in a region in Sweden ranged from 0.10 to 0.16 kg $CO_2$e kg$^{-1}$ fresh weight with 95% certainty. Their lowest estimate corresponds to the estimates for Canada.

### 3.3.3   Carbon Footprints of Livestock Products

Much progress has been achieved in reducing the carbon footprints of livestock products in Canada during the last 30 years; however, a stabilization of the emission intensities has been noticed during the last 10 years (Fig. 9). The largest reductions have been about one-third for beef and pork, but reductions have been observed for all commodities. The changes over time are due to improvements in management practices, and more productive crop varieties and livestock breeds. The main driver of the decrease is the substantial increase in production such as better crop yields, higher milk production and improved rates of weight gain. In this example, the changes in soil carbon were not included. It should be mentioned that estimates prior to 2001 are more uncertain than in recent years because a greater number of assumptions had to be made since critical data sets were unavailable such as the provincial scale animal diet by livestock category.

As seen in Fig. 10, even though GHG emissions per cow have increased from 120 to 172 kg $CH_4$ $hd^{-1}$ $year^{-1}$, the GHG emissions per liter of milk have decreased from 1.2 to 1.0 kg $CO_2e$ $kg^{-1}$ during the last 30 years. Hence, we now need less cows to produce the same amount of milk.

It is important that carbon footprint calculations take into account that some industries produce more than one product. For example, the dairy industry produces milk and meat. Because this industry is very important in Eastern Canada and not as important in Western Canada, Vergé et al. (2018) reported a 22% lower carbon footprint values for the meat produced in Eastern Canada as compared to Western Canada. In fact, beef cattle produce other coproducts that could also be used to reduce the environmental burden of producing meat. Using mass and economic allocation factors

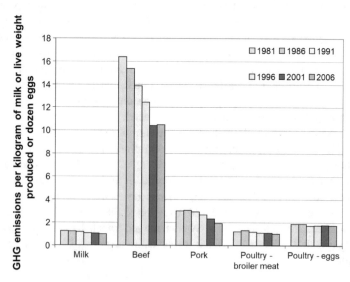

**Fig. 9**  Greenhouse gas emissions per kg of milk or live weight or dozen eggs in Canada, 1981–2006

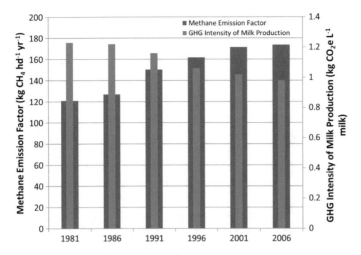

**Fig. 10** Comparison over time between the CH$_4$ emissions per cow and the GHG emissions per liter of milk in Canada, 1981–2006

**Table 5** Slaughtering mass balance and coproduct allocation factors of beef cattle in Canada, 2006

| | Mass balance | Coproduct allocation factors | |
| --- | --- | --- | --- |
| | (% SLW) | Mass (%) | Economic (%) |
| Wastes | 21.3 | | |
| Meat, primal cuts | 44.9 | 57.1 | 86.9 |
| Render products | 25.7 | 32.6 | 4.7 |
| Hide, raw | 4.9 | 6.2 | 6.0 |
| Offal | 3.2 | 4.1 | 2.4 |
| Total | 100 | 100 | 100 |

*Source* Desjardins et al. (2012)

to distribute the environmental burden, Desjardins et al. (2012) showed that at the exit of the slaughterhouse, from a mass and economic perspective, the allocation of the environmental burden was very different (Table 5). If we are more interested in the economic value of the coproducts than in their weight, the numbers from an economic perspective are then more meaningful than they are from a mass perspective. However, if we want to promote good environmental management, as discussed in Vergé et al. (2016) the mass allocation approach should be used.

### 3.3.4 Carbon Footprints Per Unit of Protein

To compare different food products, a common performance measure must be used. Proteins are essential nutrients for the human body and also a common component

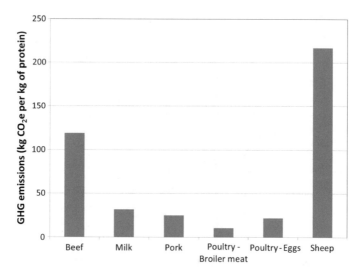

**Fig. 11** Carbon footprints per unit of protein for livestock products in Canada during 2001 (Dyer et al. 2010a)

of livestock production. Therefore, proteins provide a useful unit of comparison (discussed in Sect. 3.4). Figure 11 shows the GHG emissions associated with the protein production of several livestock products in Canada in 2006 (Dyer and Vergé 2015; Dyer et al. 2010a). These studies showed that certain sources of proteins result in more GHG emissions than other sources. Because of the relatively low fecundity of beef cattle and sheep and because they produce large amount of $CH_4$ during feed digestion (Dyer et al. 2014c; Vergé et al. 2008), the amount of GHGs emitted per unit of protein is 10–20 times larger from these industries than from poultry (Vergé et al. 2009a). Although beef and sheep compare poorly to other industries, there are other factors to consider, such as that ruminants can convert low-quality grass-based feed to complete proteins.

The range of GHG emissions associated with protein production for ruminants, non-ruminants and soybeans and other legumes (Table 6) presented by Dyer and Vergé (2015) showed just how much the production of pulses and soybeans is far less GHG-intensive methods of producing proteins as compared with ruminants and non-ruminant livestock. Because of a generally wetter climate and greater $N_2O$ emissions, more GHGs per hectare are produced in Eastern Canada than in Western Canada (Rochette et al. 2008). However, since production systems are more intensive in Eastern Canada and more extensive in Western Canada, more proteins per hectare are produced in Eastern Canada than in Western Canada. As a result, the tons of GHG emissions per ton of protein were less in Eastern Canada than Western Canada. Based on the results in Fig. 11 and Table 6, it is clear that since certain sources of proteins result in less GHG emissions than other sources, there are opportunities to reduce GHG emissions from the agriculture sector by influencing the type of food consumed. It is also very likely that a research group such as Protein Industries Canada, which

**Table 6** GHG production in eastern and Western Canada in 2006 from animal and plant protein sources

| Source of protein | t $CO_2e$ ha$^{-1}$ | | kg protein ha$^{-1}$ | | t $CO_2e$ t$^{-1}$ protein | |
|---|---|---|---|---|---|---|
| Animal | East | West | East | West | East | West |
| Ruminants | 15.17 | 11.33 | 263 | 103 | 57.77 | 109.83 |
| Non-ruminants | 3.13 | 1.82 | 167 | 83 | 18.79 | 21.97 |
| Plant | | | | | | |
| Soybeans | 0.30 | 0.26 | 1077 | 630 | 0.28 | 0.42 |
| Other legumes | 0.41 | 0.34 | 207 | 139 | 1.98 | 2.46 |

*Source* Dyer and Vergé (2015)

is part of an alliance of over 120 stakeholders who are attempting to increase the potential of plant-based proteins for canola, peas, hemp and flax, will come up with scenarios to improve the value of plant-based proteins and coproducts.

### 3.3.5 Carbon Footprints of Milk Products

Vergé et al. (2013) compared the carbon footprints of milk and yogurt at the exit gate of processing plants in Canada for 2006. The estimates were obtained using the Canadian Food Carbon Footprint (Cafoo$^2$) calculator. The emission intensities due to farm production, transportation, processing and packaging are given (Table 7). The GHG emissions associated with the on-farm emissions account for more than 70% of the emissions for both products. The processing details for milk and yogurt are also presented. The Cafoo$^2$ calculator has been used to estimate the emission intensity for 11 dairy products by region and by the province for all of Canada (Vergé et al. 2013).

### 3.3.6 Carbon Footprints of Crops for Bioenergy Production

The increasing demand for the agriculture sector, to not only produce food but also to provide biofuels, points to the need for more information on the carbon footprints of potential feedstock crops. The most important biofuel feedstock crop in Canada is presently canola, due to the European market as a biodiesel feedstock that has grown rapidly in the last few decades. Recently, the EU Directive on the promotion of the use of energy from renewable energy stated that by 2020, 20% of the European Community gross consumption of energy had to come from renewable sources and 10% of the transport energy use needed to be from renewable sources (European Union 2009). It is expected that biofuels and bioliquids will be used extensively to meet this target. Crops without proof that their carbon footprint meets the sustainability criteria are not going to be counted toward the European Community quota fulfill-

**Table 7** Emission intensities associated with fluid milk and yogurt production in 2006

| Step | Fluid milk | Yogurt |
|---|---|---|
| | kg $CO_2$e $kg^{-1}$ | kg $CO_2$e $kg^{-1}$ |
| Farm production | 0.87 | 1.23 |
| Transportation | 0.01 | 0.02 |
| Processing | 0.07 | 0.29 |
| Packaging | 0.06 | 0.18 |
| Total | 1.00 | 1.75 |
| Processing details | kg $CO_2$e $kg^{-1}$ | kg $CO_2$e $kg^{-1}$ |
| Electricity | 0.02 | 0.12 |
| Fossil fuels | 0.04 | 0.17 |
| Water and wastewater | 0.00 | 0.00 |
| Cleaners | 0.00 | 0.00 |
| Refrigerants | 0.00 | 0.00 |
| Total | 0.07 | 0.29 |

ment. As of January 2018, biofuels had to achieve 50% GHG emission savings. For installations after October 2015, a GHG saving of at least 60% was required. In the percentage, cultivation of the canola, processing and transportation and distribution must be included. The fossil reference is 83.8 g $CO_2$e $MJ^{-1}$. Canada and many countries have prepared reports detailing the regional GHG emissions associated with the cultivation of canola. GHG emission estimates for various regions vary due to the production practices, soil and climate. GHG emissions associated with seeding, fertilizer production, $N_2O$ field emissions, pesticide production and field operations all need to be included.

## 3.4 GHG Emission Indicator Suite

To make the link between the GHG emission intensity estimates provided by the research discussed above and the government policies that can best mitigate agricultural GHG emissions, the agricultural carbon footprints must be represented by an effective, quantitative suite of indicators. Canada has a long history of developing agro-environmental indicators (AEIs) (Huffman et al. 2000). Under this national framework, Desjardins et al. (2010b) and Worth et al. (2016) provided AEIs for $N_2O$, $CO_2$ and $CH_4$. Policy-relevant indicators provide a metric against which public policy issues can be measured (Hammond et al. 1995). The framework outlined by OECD (2008) and followed in Canada by McRae et al. (2000) included policy relevance, analytical soundness (scientific rigor and defensibility), measurability and ease of interpretation.

The most basic agricultural GHG indicator is the GHG emission intensities on an area basis. Area-based intensities are an improvement on emission totals when comparing total GHG emissions from specific crops or crop groups, such as fossil $CO_2$ emissions from small grain cereals (Dyer and Desjardins 2005a). However, land-based intensities are inadequate and misleading when crops such as perennial forages and annual field crops that are so fundamentally different are compared. A more integrative approach is to assess the GHG emission budgets for individual farms through the whole-farm concept (Janzen et al. 2006; Kröbel et al. 2013), which provides useful guidance to managers of those specific farms. But the transition from guiding farm decisions to measuring public policy requires a suite of indicators assembled from meaningful product groups, farm-type GHG emission budgets and land-use categories.

In their assessment of potential GHG indicators, Dyer et al. (2018) explored interactions between GHG sources and three farm product groups defined by their nutritional components: animal-equivalent protein (AeP) composed of all the essential amino acids needed in the human diet (Tessari et al. 2016), vegetable oils (Voil) and carbohydrates (CHO). Since not all CHO are suitable for human consumption, they were separated into two groups: bread quality grains (CHO bread) and feed grains (CHO coarse). Voil crops provide both cooking oils and biodiesel feedstock, thus reducing fossil diesel consumption. Canola, which is the most important Voil source in Canada and a major biodiesel feedstock, can offset a fraction of the global wildlife habitat and biodiversity losses associated with using palm oil as biodiesel feedstock (Dyer et al. 2011c). Canola has undergone dramatic growth in Western Canada, largely filling the gap from the decline in summer fallow (Shrestha et al. 2014).

The GHG protein indicator has been used for comparing GHG emission intensities for multiple agricultural industries and commodities (Dyer and Vergé 2015; Dyer et al. 2010a), and the policy relevance of this indicator has been effectively demonstrated (Dyer et al. 2011d, 2013, 2015a, b; Vergé et al. 2012). An alternative to AeP, live weight (LW), may be a suitable replacement for AeP when the product is the livestock carcass (Vergé et al. 2008, 2009a, b). But LW does not effectively represent milk and egg production, nor does LW facilitate inter-commodity comparisons of livestock types. Not only is AeP suitable for inter-commodity comparison, but it allowed comparison of all livestock products with pulse-based proteins. The rapid increase in global demand for protein reflects the rise in income among developing nations and that protein is an essential dietary component of the human diet (Dyer and Vergé 2015). Hence, this dietary imperative means that access to adequate protein is as critical to the global community as access to food in general.

Dyer et al. (2018) grouped agricultural GHG emissions as: all agricultural activities (All); all livestock (LS); ruminant livestock (Rum); and non-ruminant livestock (Non-r). The hierarchy among the GHG sources is that the LS GHG emission term is part of the All GHG emission term, but this term is also the sum of the Rum and Non-r GHG emissions. Quantifying the All category of GHG emissions was the only GHG term that cannot be fully quantified by ULICEES. They also explored the relationships among the three main land-use categories and four crop groups, comprising forages, grains, pulses and oilseeds. The three land-use categories were food grains and oilseeds (FGO), the livestock crop complex (LCC, described above) and the non-livestock residual (NLR), the quantity that is in excess of (not required in) the livestock diet. Perennial forages are always part of the LCC because they were all consumed by ruminant livestock. Similarly, oilseeds were always part of the FGO because of their main product being vegetable oil, even though their meal fraction may be used as livestock feed.

Dyer et al. (2018) defined two groups of indicators. One group of indicators were defined as ratios of plant products to the All GHG emission term. A second group was defined as ratios of protein to GHG emissions from either the Non-r or the Rum GHG sources. For the second, the Rum GHG rates per unit of protein were three times as high as the Non-r rates per unit of protein in Eastern Canada and five times as high for Western Canada. For the first group of indicators, it was difficult to make meaningful east–west comparisons for CHO bread and Voil-based indicators because so little of these two product groups were produced in Eastern Canada. For the CHO coarse indicator, the east–west differences were small since they both reflected livestock production more than crop production. It was not appropriate to report the quantitative results from Dyer et al. (2018) here because that assessment used the All GHG emission term. They also demonstrated that the first group of indicators could be used in the reciprocal form as a means of comparing the GHG efficiencies of crop areas, as well as their emission intensities. While ULICEES played a major role in this analysis, it also demonstrated the need for sound GHG emission estimates from all agricultural commodities.

## 4   Reducing GHG Emissions and Energy Production

There are many different options for reducing on-farm GHG emissions (Kröbel et al. 2013). Livestock management practices such as different winter feeding strategies, including swath grazing and bale grazing, are good examples. In the case of swath grazing, the last cut of hay or a late-seeded cereal crop is left in swath from which the cattle self-feed during the winter. In the case of bale grazing, bales or hay, often with bales of straw, is distributed over the field from which the cattle feed. Both of these practices result in less manure handling, less equipment use, less fuel cost and less associated GHG emissions. However, in order that the impact of these practices be recognized, it is important that the impacts are well understood and that the benefits be counted in the GHG inventories (VanderZaag et al. 2013). This can only be done if the activity level of these practices is known at the national scale.

The food choice that consumers make can also help the sector reduce its GHG emissions. For example, Vergé et al. (2012) demonstrated that a 10% shift from beef to pork production in Canada would result in a reduction of 2 Mt $CO_2$e year$^{-1}$ which is equivalent to a 3% reduction of GHG emissions from the agriculture sector. With the carbon footprint data available, it is then possible to examine the impact of many other potential scenarios. However, the biggest gains are likely to be from a shift from meat proteins to plant proteins where gains of the order of 10–100 times are possible (Dyer and Vergé 2015). We have also shown that depending on the allocation of environmental burden the carbon footprint values can be very different. The carbon footprint of agricultural products will continue to decrease, not only because of an increase in production efficiencies but also because, as coproducts become valuable, they can then share in the environmental burden.

Agricultural production in Canada will likely be influenced as much by climate change adaptation, as by the obligation to reduce Canada's GHG emissions. Adaptation will take place in response to both the new agro-climates caused by climate change and stronger market opportunities for food production. To offset expected declines in red meat consumption, for example, growth is most likely to occur in the horticulture and greenhouse industries, particularly in the warmest regions including the Great Lakes Basin and Saint Lawrence Lowlands, and the Maritimes and Southern British Columbia (Agriculture and Agri-Food Canada (AAFC) 2016; Statistics Canada 2016). The carbon footprints of these industries are complicated by the diversity of their food products. These carbon footprint calculations are driven by energy consumption and inputs such as N fertilizers. The energy use calculations are embedded in two spreadsheet models (Dyer and Desjardins 2018; Dyer et al. 2011c). When augmented by the $N_2O$ emission estimates from the N fertilizer use for these intensive crops, these data can be an important policy tool in developing a comprehensive land-use strategy in our rapidly changing world. A key problem that we anticipate for the agricultural sector is that production systems are sometimes not sufficiently diversified to be resilient when faced with anticipated climate extremes. This problem will be exacerbated by the need to feed a world population that is increasing both in numbers and wealth. Economies of scale and the need to maximize production have led to increased dependence on a limited number of high yielding crops, and the concentration and specialization of animal production systems. Both of these examples tend to increase vulnerability to extreme weather, as well as to pests and diseases. An integrated approach is then needed within all production systems in order to build resiliency, adapt to risk posed by climate change and reduce threats to food security.

Scientists need to anticipate environmental problems and provide information to policy makers, particularly robust, policy-relevant GHG indicators, so they can develop a good understanding of the science and propose relevant policies. There is a need to bridge the communication gaps between these two groups by demonstrating how mankind, including agriculture, is changing the climate and what policies could help minimize and manage environmental risks. The translation of the science and communication of science beyond publishing in scientific journals is essential for bridging the gap. The science community needs to assist government and community spokespersons with statements that are easily understood by the general public. This will greatly encourage public support of adaptation and mitigation policies and practice and help the Government of Canada and the agriculture sector in achieving their goal of reducing GHG emissions.

There are two relevant questions about biofuels to the sector's carbon footprint. The first is whether the energy derived from the biofuels exceeds the total energy used to produce the feedstock crops. The second is whether the net GHG emission reductions from the biofuel feedstock crops justify the diversion of cropland from food production. These two questions have been addressed in several analyses (Dyer et al. 2010b, 2011a; Dyer and Vergé 2015). As discussed above, canola, as a biodiesel feedstock, also displaces palm oil and mitigates the impact of that crop on tropical ecosystems, which adds justification for canola that goes beyond the carbon footprint. Where dislocation of livestock is a possible outcome of the expansion of biofuel feedstock production, the carbon footprint will extend beyond the cultivation of the feedstock crop. The ability of biofuels to reduce GHG emissions depends on either previous or alternative uses of the land targeted for feedstock production. Although their initial biodiesel feedstock analysis only dealt with crop-type interactions, Dyer et al. (2010b) cautioned that some impact of feedstock production on the carbon footprint of livestock industries would likely be unavoidable. Dyer et al. (2011c) found that for the expansion of feedstock crops into land that supports non-ruminant livestock (poultry and pork), the impact would be straightforward since there is no significant fallback on grazing. For ruminants, however, these interactions are highly complex, even when considered on the one-dimensional basis of GHG emissions taken in this analysis. Farmers with ruminants can respond to reduced feed grain supply in two ways: by either reducing their livestock numbers or by returning to a more roughage-based diet with more forage and less grain. Dyer et al. (2011c) found that the displacement of ruminants by biofuel feedstock is an effective GHG reduction strategy if the populations of those displaced animals are actually reduced. However, when they are simply transferred to the more forage-based diet, the enhanced benefit from reduced enteric methane emissions is either canceled out or reversed. Dyer et al. (2011c) did not deal with the changes in soil carbon as a result of land-use changes nor take into account that the sector would produce less meat.

In a similar follow-up analysis, Dyer et al. (2015b) dealt specifically with the impact of expanded canola for biofuel feedstock on the Prairie beef industry. While the only crops directly displaced by the assumed canola expansion were the feed grains, this analysis accounted for the increase in forage areas due to the assumed beef diet. However, this follow-up analysis ensured a constant meat supply with the assumption that total protein supply must be maintained. This analysis also accounted for sequestering of atmospheric $CO_2$ as soil carbon. It considered two scenarios: (1) relocating the displaced feedlot cattle to pasture and rangeland, and a diet much richer in hay, and (2) slaughtering the cattle destined for finishing in feedlots as veal (no finishing). The net carbon footprint of the expanded canola exceeded the fossil $CO_2$ emission of the equivalent energy quantity of petrodiesel by 16% in the dislocation scenario and was exceeded by the fossil $CO_2$ emission offsets by 32% in the slaughter scenario 4. Supported by these results, Dyer et al. (2015b) concluded that the expansion of canola for biodiesel feedstock is unlikely to be sustainable if ruminant livestock are displaced into a more forage-dependant production system by the expansion.

To our knowledge, there has not been an analysis of how Canadian biofuel production affects indirect land-use change outside of Canada, although studies for biofuel production in other countries suggest this could be significant (Acquaye et al. 2011;

Hansen et al. 2014; Reinhard and Zah 2011) albeit uncertain (Flysjö et al. 2012; Mathews and Tan 2009).

The use of farm animal waste for biogas generation is also an excellent way where producers can reduce their on-farm GHG emissions while generating energy. This additional source of revenue can help make their farm more sustainable. However, Flesch et al. (2011) showed that the GHG reduction associated with a biodigester can sometimes be substantially reduced by fugitive emissions. Biodigesters are therefore an important source of energy, but they also produce digestates which provide soil with organic content, as well as undigested fibers that can be used as bedding for the animals (Guest et al. 2017).

# 5 Conclusions

At the start of this chapter, we described the techniques used for quantifying the GHG emissions from agricultural sources. Accurate estimates of agricultural GHG emissions are an important part of the Canadian GHG mitigation efforts for two reasons. First, as a developed nation with an intensive agricultural industry, Canada must contribute to the international body of science about quantifying GHG emissions, rather than relying on the research conducted in other countries. Second, it is essential that any impacts from field conditions on the agricultural emission GHG budget that are unique to Canada need to be quantified and incorporated into Canadian GHG emission models, indicators and carbon footprint reporting schemes. We briefly described some of the models used to quantify GHG emissions from the agriculture sector. We reported that the sector accounts for about 12% of the GHG emissions in Canada. We gave the percentage contributions of the main agricultural sources of $CO_2$, $N_2O$ and $CH_4$ and the carbon footprint values for the major crops and livestock products in Canada. We presented estimates of the carbon footprints of agricultural products from other research groups. This is, however, the first publication which presents carbon footprint estimates of most of the agricultural products for a country. We reported that the carbon footprint value per unit of proteins for soybeans is a factor of hundred less than for ruminants. Using barley as an example, we showed that there can be substantial regional differences in the carbon footprint values of an agricultural product. We discussed how the carbon footprint values for canola are being used for determining if the production of canola in certain regions meets the requirement of the European Community to be sold for producing biodiesel. We gave several examples where the sharing of the environmental burden with coproducts helped reduce the carbon footprint of the primary product. The carbon footprint information and emission indicators presented above should provide consumers the opportunity to make food choices that can reduce GHG emissions from the agriculture sector. We briefly presented a tool which can help producers reduce their on-farm GHG emissions. We provided information that should help policy makers formulate environmentally sustainable policies that would reduce the impact of the agriculture sector on climate change. It is fairly clear that more information on the accuracy of these estimates is required. It must also be made clear that, if applicable, soil carbon change and indirect land-use changes should be included in the carbon footprint calculations.

# References

Acquaye AA et al (2011) Identification of 'carbon hot-spots' and quantification of GHG intensities in the biodiesel supply chain using hybrid LCA and structural path analysis. Environ Sci Technol 45(6):2471–2478

Agriculture and Agri-Food Canada (AAFC) (2016) Statistical overview of the canadian vegetable industry—2015. Agriculture and Agri-Food Canada. Retrieved from http://www.agr.gc.ca/eng/industry-markets-and-trade/canadian-agri-food-sector-intelligence/horticulture/horticulture-sector-reports/statistical-overview-of-the-canadian-vegetable-industry-2015/?id=1478646189894#a1.1

Baldocchi D et al (2001) FLUXNET: a new tool to study the temporal and spatial variability of ecosystem-scale carbon dioxide, water vapor, and energy flux densities. Bull Am Meteor Soc 82(11):2415–2434

Campbell CA et al (2000) Organic C accumulation in soil over 30 years in semiarid southwestern Saskatchewan—effect of crop rotations and fertilizers. Can J Soil Sci 80:179–192

Clark M, Tilman D (2017) Comparative analysis of environmental impacts of agricultural production systems, agricultural input efficiency, and food choice. Environ Res Lett 12(6):064016

Cole CV et al (1997) Global estimates of potential mitigation of greenhouse gas emissions by agriculture. Nutr Cycl Agroecosyst 49(1–3):221–228

Desjardins RL et al (2001) Canadian greenhouse gas mitigation options in agriculture. Nutr Cycl Agroecosyst 60(1):317–326

Desjardins RL et al (2004) Evaluation of a micrometeorological mass balance method employing an open-path laser for measuring methane emissions. Atmos Environ 38:6855–6866

Desjardins RL et al (2010a) Multiscale estimates of $N_2O$ emissions from agricultural lands. Agric For Meteorol 150:817–824

Desjardins RL et al (2010b) Chapter 16: agricultural greenhouse gases. In: Eilers, W, Mackay R, Graham L, Lefebvre A (eds) Environmental sustainability of Canadian agriculture. Agri-environmental indicator report series, report #3. Agriculture and Agri-Food Canada, Ottawa, ON

Desjardins RL, Smith WN, Grant BB, Campbell C, Riznek R (2005) Management strategies to sequester carbon in agricultural soils and to mitigate greenhouse gas emissions. Clim Change 70:283–297

Desjardins RL, Worth DE, MacPherson JI, Mauder M, Bange J (2019) Aircraft-based density measurements. In: Foken T (ed) Springer handbook of atmospheric measurements. Springer, Boulder

Desjardins RL et al (2012) The carbon footprint of beef production. Sustainability 4:3279–3301

Desjardins RL et al (2016) Greenhouse Gas Emission Intensities of Agricultural Products. In: Clearwater RL, Martin T, Hoppe T (eds) Environmental sustainability of Canadian agriculture: agri-environmental indicator report series—report #4. Agriculture and Agri-Food Canada, Ottawa, ON

Desjardins RL et al (2018) The challenge of reconciling bottom-up agricultural methane emissions inventories with top-down measurements. Agric For Meteorol 248:48–59

Dyer JA, Desjardins RL (2003) Simulated farm fieldwork, energy consumption and related greenhouse gas emissions in Canada. Biosys Eng 85(4):505–513

Dyer JA, Desjardins RL (2005a) Analysis of trends in $CO_2$- emissions from fossil fuel use for farm fieldwork related to harvesting annual crops and hay, changing tillage practices and reduced summer fallow in Canada. J Sustain Agric 25(3):141–155

Dyer JA, Desjardins RL (2005b) A simple meta-model for assessing the contribution of liquid fossil fuel for on-farm fieldwork to agricultural greenhouse gases in Canada. J Sustain Agric 27(1):71–90

Dyer JA, Desjardins RL (2009) A review and evaluation of fossil energy and carbon dioxide emissions in Canadian agriculture. J Sustain Agric 33:210–228

Dyer JA, Desjardins RL (2018) Energy use and fossil $CO_2$ emissions for the Canadian fruit and vegetable industries. Energy Sustain Dev 47:23–33

Dyer JA, Desjardins RL, McConkey BG (2014a) Assessment of the carbon and non-carbon footprint interactions of livestock production in Eastern and Western Canada. Agroecol Sustain Food Syst 38:541–572

Dyer JA, Desjardins RL, McConkey BG (2014b) The fossil energy use and $CO_2$ emissions budget for Canadian agriculture. In: Bundschuh J (ed) Sustainable energy solutions. Sustainable energy developments. Taylor and Francis/CRC Press, Boca Raton, pp 77–96

Dyer JA, Vergé XPC, Desjardins RL, Worth DE (2014c) A comparison of the greenhouse gas emissions from the sheep industry with beef production in Canada. Sustain Agric Res 3(3):65–75

Dyer JA, Desjardins RL, McConkey BG, Kulshreshtha SN, Vergé XPC (2013) Integration of farm fossil fuel use with local scale assessments of biofuel feedstock production in Canada. In: Fang Z (ed) Biofuel economy, environment and sustainability. InTech Open Access Publisher, Rijeka, Croatia, pp 97–122

Dyer JA, Hendrickson OQ, Desjardins RL, Andrachuk HL (2011a) An environmental impact assessment of biofuel feedstock production on agro-ecosystem biodiversity in Canada. In: Contreras LM (ed) Agricultural policies: new developments. Nova Science Publishers, Hauppauge, NY, p 281

Dyer JA, Kulshreshtha SN, McConkey BG, Desjardins RL (2011b) An assessment of fossil fuel energy use and $CO_2$ emissions from farm field operations using a regional level crop and land use database for Canada. Energy 35:2261–2269

Dyer JA, Vergé XPC, Desjardins RL, McConkey BG (2011c) Implications of biofuel feedstock crops for the livestock feed industry in Canada. In: Dos Santos Bernardes MA (ed) Environmental impact of biofuels. InTech Open Access Publisher, Rijeka, Croatia, pp 161–178

Dyer JA, Vergé XPC, Kulshreshtha SN, Desjardins RL, McConkey BG (2011d) Areas and greenhouse gas emissions from feed crops not used in Canadian livestock production in 2001. J Sustain Agric 35:780–803

Dyer JA, Vergé XPC (2015) The role of Canadian agriculture in meeting increased global protein demand with low carbon emitting production. Agronomy 5:569–586

Dyer JA, Vergé XPC, Desjardins RL, Worth D (2010a) The protein-based GHG emission intensity for livestock products in Canada. J Sustain Agric 34(6):618–629

Dyer JA, Verge XPC, Desjardins RL, Worth DE, McConkey BG (2010b) The impact of increased biodiesel production on the greenhouse gas emissions from field crops in Canada. Energy Sustain Dev 14:73–82

Dyer JA, Vergé XPC, Desjardins RL, Worth DE (2015a) An assessment of greenhouse gas emissions from co-grazing sheep and beef in Western Canadian rangeland. In: McHenry MP, Kulshreshtha SN, Lac S (eds), Agriculture management for climate change. Nova Science Publishers, Inc., pp 13–29

Dyer JA, Vergé XPC, Desjardins RL, Worth DE (2015b) Changes in greenhouse gas emissions from displacing cattle for biodiesel feedstock. In: Biernat K (ed) Biofuels—Status and perspective. InTech Open Access Publisher, Rijeka, Croatia, pp 351–375

Dyer JA, Vergé XPC, Desjardins RL, Worth DE (2018) District scale GHG emission indicators for Canadian field crop and livestock production. Agronomy 8(9):190

Ellis JL et al (2009) Modeling methane production from beef cattle using linear and nonlinear approaches. J Anim Sci 87(4):1334–1345

Environment and Climate Change Canada (2017) National inventory report 1990–2015: greenhouse gas sources and sinks in Canada. Part 3. Environment and climate change Canada. Pollutant Inventories and Reporting Division, Gatineau, QC

European Union (2009) Directive 2009/28/EC of the European Parliament and of the Council on the promotion of the use of energy from renewable sources and amending and subsequently repealing directives 2001/77/EC and 2003/30/EC. In: European Union (ed) Official Journal of the European Union

Flesch TK, Desjardins RL, Worth DE (2011) Fugitive methane emissions from an agricultural biodigester. Biomass Bioenerg 35:3927–3935

Flysjö A, Cederberg C, Henriksson M, Ledgard S (2012) The interaction between milk and beef production and emissions from land use change—critical considerations in life cycle assessment and carbon footprint studies of milk. J Clean Prod 28:134–142

Gan Y et al (2014) Improving farming practices reduces the carbon footprint of spring wheat production. Nat Commun 5:5012

Goglio P et al (2015) Accounting for soil carbon changes in agricultural life cycle assessment (LCA): a review. J Clean Prod 104(1):23–39

Grant BB, Smith WN, Desjardins RL, Lemke RL, Li C (2004) Estimated $N_2O$ and $CO_2$ emissions as influenced by agricultural practices in Canada. Clim Change 65:315–332

Guest G et al (2017) A comparative life cycle assessment highlighting the trade-offs of a liquid manure separator-composter in a Canadian dairy farm. J Clean Prod 143:824–835

Hammond A, Adriaanse A, Bodenburg E, Bryand D, Woodward R (1995) Environmental indicators: a systematic approach to measuring and reporting on environmental policy performance in the context of sustainable development. World Resources Institute

Hansen J, Ruedy R, Sato M, Lo K (2010) Global surface temperature change. Rev Geophys 48:RG4004

Hansen SB, Olsen SI, Ujang Z (2014) Carbon balance impacts of land use changes related to the life cycle of Malaysian palm oil-derived biodiesel. Int J Life Cycle Assess 19(3):558–566

Hillier J, Hawes C, Squire G, Hilton A, Wale S (2009) The carbon footprints of food crop production. Int J Agric Sustain 7(2):107–118

Hörtenhuber S, Piringer G, Zollitsch W, Lindenthal T, Winiwarter W (2014) Land use and land use change in agricultural LCAs and carbon footprints—the case for regionally specific LUC versus other methods. J Clean Prod 73:31–39

Huffman E, Eilers RG, Padbury G, Wall G, MacDonald KB (2000) Canadian agri-environmental indicators related to land quality: integrating census and biophysical data to estimate soil cover, wind erosion and soil salinity. Agr Ecosyst Environ 81(2):113–123

IPCC (2007) Climate change 2007: the physical science basis. Contribution of working group I to the fourth assessment report of the intergovernmental panel on climate change. Cambridge University Press, Cambridge, UK and New York, USA, 996 pp

IPCC (2013) Climate change 2013: the physical science basis. Contribution of working group I to the fifth assessment report of the intergovernmental panel on climate change. Cambridge University Press, Cambridge, UK and New York, NY USA

Janzen HH et al (2006) A proposed approach to estimate and reduce net greenhouse gas emissions from whole farms. Can J Soil Sci 86:401–416

Janzen HH, Desjardins RL, Asseline JMR, Grace B (1999) The health of our air: toward sustainable agriculture in Canada. Research Branch Agriculture and Agri-Food Canada, Ottawa, Ontario

Karimi-Zindashty Y et al (2012) Sources of uncertainty in the IPCC Tier 2 Canadian livestock model. J Agric Sci 150:556–569

Kröbel R et al (2013) A proposed approach to estimate and reduce the environmental impact from whole farms. Acta Agric Scand Sect A Anim Sci 62(4):225–232

Lal R, Bruce JP (1999) The potential of world croplands soils to sequester C and mitigate the greenhouse effect. Environ Sci Policy 2(2):177–185

Li C et al (2012) Manure-DNDC: a biogeochemical process model for quantifying greenhouse gas and ammonia emissions from livestock manure systems. Nutr Cycl Agroecosyst 93(2):163–200

Liang BC, Aruna H, Ma BL (2016) Carbon footprints in crop rotation systems. In: Ma BL (ed) Crop rotations farming practices, monitoring and environmental benefits. Nova Science Publishers, New York, NY, pp 157–176

Lindorfer J, Fazeni K, Steinmüller H (2014) Life cycle analysis and soil organic carbon balance as methods for assessing the ecological sustainability of 2nd generation biofuel feedstock. Sustain Energy Technol Assess 5:95–105

Mathews JA, Tan H (2009) Indirect land use emissions in the life cycle of biofuels: the debate continues. Biofuels Bioprod Biorefin 3(3):305–317

McRae T, Smith CAS, Gregorich EG (eds) (2000) Environmental sustainability of Canadian agriculture. Report of the agri-environmental indicator project. Agriculture and Agri-Food Canada, Ottawa, Canada, 224 pp

Müller-Wenk R, Brandão M (2010) Climatic impact of land use in LCA-carbon transfers between vegetation/soil and air. Int J Life Cycle Assess 15(2):172–182

Nagy N (2000) Energy and greenhouse gas coefficients inputs used in agriculture. Report to the Prairie Adaptation Research Collaborative (PARC). University of Saskatchewan, Saskatoon, Canada

OECD (2008) Environmental performance of agriculture in OECD countries since 1990. Organization for Economic Cooperation and Development

Pattey E et al (2006a) Towards standards for measuring greenhouse gas fluxes from agricultural fields using instrumented towers. Can J Soil Sci 86(3):373–400

Pattey E et al (2006b) Application of a tunable diode laser to the measurement of $CH_4$ and $N_2O$ fluxes from field to landscape scale using several micrometeorological techniques. Agric For Meteorol 136(3–4):222–236

Pattey E et al (2007) Tools for quantifying $N_2O$ emissions from agroecosystems. Agric For Meteorol 142:103–119

Pradhan P, Reusser DE, Kropp JP (2013) Embodied greenhouse gas emissions in diets. PLoS ONE 8(5):e62228

Reinhard J, Zah R (2011) Consequential life cycle assessment of the environmental impacts of an increased rapemethylester (RME) production in Switzerland. Biomass Bioenerg 32(6):2361–2373

Rochette P, Eriksen-Hamel N (2008) Chamber measurements of soil nitrous oxide flux: are absolute values reliable? Soil Sci Soc Am J 72:331–342

Rochette P et al (2008) Estimation of $N_2O$ emissions from agricultural soils in Canada. I—development of a country specific methodology. Can J Soil Sci 88:641–654

Rogelj J et al (2016) Paris agreement climate proposals need a boost to keep warming well below 2 °C. Nature 534:631–639

Röös E, Sundberg C, Hansson P-A (2010) Uncertainties in the carbon footprint of food products: a case study on table potatoes. Int J Life Cycle Assess 15(5):478–488

Ruddiman WF (2003) The anthropogenic greenhouse era began thousands of years ago. Clim Change 61:261–293

Schmidinger K, Stehfest E (2012) Including $CO_2$ implications of land occupation in LCAs—method and example for livestock products. Int J Life Cycle Assess 17(8):962–972

Shrestha BM et al (2014) Change in carbon footprint of canola production in the Canadian Prairies from 1986 to 2006. Renew Energy 63:634–641

Smith WN, Desjardins RL, Grant BB (2001) Estimated changes in soil carbon associated with agricultural practices in Canada. Can J Soil Sci 81:221–227

Smith WN, Grant BB, Desjardins RL, Lemke RL, Li C (2004) Estimates of the interannual variations of $N_2O$ emissions from agricultural soils in Canada. Nutr Cycl Agroecosyst 68:37–45

Smith WN et al (1997) The rate of carbon change in agricultural soils at the landscape level. Can J Soil Sci 77(2):219–229

Snyder CS, Bruulsema TW, Jensen TL (2007) Greenhouse gas emissions from cropping systems and the influence of fertilizer management—a literature review. International Plant Nutrition Institute, Norcross, Georgia

Statistics Canada (2016) Farms, by farm type and by province, 2011. Statistics Canada. Retrieved from https://www150.statcan.gc.ca/n1/pub/11-402-x/2012000/chap/ag/tbl/tbl07-eng.htm

Statistics Canada (2018a) Table: 32-10-0406-01. Land Use. Statistics Canada

Statistics Canada (2018b) Table: 32-10-0424-01. Cattle and calves on census day. Statistics Canada

Statistics Canada (2018c) Table: 32-10-0425-01. Sheep and lambs on census day. Statistics Canada

Statistics Canada (2018d) Table: 32-10-0426-01. Pigs on census day. Statistics Canada

Statistics Canada (2018e) Table: 32-10-0427-01. Other livestock on census day. Statistics Canada

Statistics Canada (2018f) Table: 32-10-0428-01. Poultry inventory on census day. Statistics Canada

Stossel F, Juraske R, Pfister S, Hellweg S (2012) Life cycle inventory and carbon and water FoodPrint of fruits and vegetables: application to a Swiss retailer. Environ Sci Technol 46:3253–3262

Tessari P, Late A, Mosca G (2016) Essential amino acids: master regulators of nutrition and environmental footprint? Sci Rep 6:26074

Tilman D, Balzer C, Hill J, Befort BL (2011) Global food demand and the sustainable intensification of agriculture. Proc Natl Acad Sci USA 108(50):20260–20264

van Soest HL et al (2017) Low-emission pathways in 11 major economies: comparison of cost-optimal pathways and Paris climate proposals. Clim Change 142:491–504

VanderZaag AC, Flesch TK, Desjardins RL, Baldé H, Wright T (2014) Methane emissions from two dairy farms: Seasonal and manuremanagement effects. Agric For Meteorol 194:259–267

VanderZaag AC, MacDonald JD, Evans L, Vergé XPC, Desjardins RL (2013) Towards an inventory of methane emissions from manure management that is responsive to changes on Canadian farms. Environ Res Lett 8(3):035008

Vergé XPC, Dyer JA, Desjardins RL, Worth D (2007) Greenhouse gas emissions from the Canadian dairy industry in 2001. Agric Syst 94(3):683–693

Vergé XPC, Dyer JA, Desjardins RL, Worth D (2008) Greenhouse gas emissions from the Canadian beef industry. Agric Syst 98:126–134

Vergé XPC, Dyer JA, Desjardins RL, Worth D (2009a) Greenhouse gas emissions from the Canadian pork industry. Livestock Sci 121:92–101

Vergé XPC, Dyer JA, Desjardins RL, Worth D (2009b) Long-term trends in the greenhouse gas emissions from the Canadian poultry industry. J Appl Poult Res 18:210–222

Vergé XPC et al (2012) A greenhouse gas and soil carbon model for estimating the carbon footprint of livestock production in Canada. Animals 2:437–454

Vergé XPC et al (2013) Carbon footprint of Canadian dairy products. Calculations and issues. J Dairy Sci 96:6091–6104

Vergé XPC, Maxime D, Desjardins RL, VanderZaag AC (2016) Allocation factors and issues in agricultural carbon footprint: a case study of the Canadian pork industry. J Clean Prod 113:587–595

Vergé XPC, VanderZaag AC, Desjardins RL, McConkey BG (2018) Synergistic effects of complementary production systems help reduce livestock environmental burdens. J Clean Prod 200:858–865

Wiedemann SG et al (2015) Resource use and greenhouse gas intensity of Australian beef production: 1981–2010. Agric Syst 133:109–118

Worth DE et al (2016) Agricultural greenhouse gases. In: Clearwater RL, Martin T, Hoppe T (eds) Environmental sustainability of Canadian agriculture: agri-environmental indicator report series—report #4. Agriculture and Agri-Food Canada, Ottawa, ON

# Extreme Inequality and Carbon Footprint of Spanish Households

Luis Antonio López, Guadalupe Arce and Mònica Serrano

**Abstract** Palma's papers (Development and Change 42:87–153, 2011, Development and Change 45:1416–1448, 2014) have shown that in most countries exist extreme inequalities in income distribution. This extreme inequality occurs because higher income population appropriates a significant part of income previously received by lower income groups without affecting middle-income group, which is relatively stable. This fact makes the gap between rich and poor wider than ever. In this chapter, we propose a measure that allows us to identify the impact that inequality in income distribution has on carbon footprint of households' consumption. Through interdecile ratio 10/1–4 of the households' footprint, it is assessed how social unsustainability (associated with the extreme inequality of income, which is transferred through the consumer spending) generates greater environmental unsustainability. We use a multi-regional input–output model to calculate the carbon footprint associated with households' consumption according to different levels of income, from the highest income level to those who are below poverty line and at risk of social exclusion. An empirical application is carried out for the carbon footprint of Spanish households from 2006 to 2013. This period allows us to assess further the impact of the financial and economic crisis started in 2008. Finally, using a regression analysis, we evaluate how changes in domestic and imported household consumption and changes in consumption inequalities (measured through Gini index and Palma ratio) determine changes in household carbon footprint considering the different

L. A. López (✉)
Faculty of Economics and Business, University of Castilla-La Mancha,
Plaza de la Universidad 1, 02071 Albacete, Spain
e-mail: Luis.LSantiago@uclm.es

G. Arce
Department of Fundations of Economic Analysis and Quantitative Economics,
Complutense University of Madrid, Campus de Somosaguas, 28223 Pozuelo de Alarcón
(Madrid), Spain
e-mail: Garce01@ucm.es

M. Serrano
Department of Economics, University of Barcelona, Av. Diagonal 696,
08034 Barcelona, Spain
e-mail: monica.serrano@ub.edu

© Springer Nature Singapore Pte Ltd. 2020                                          35
S. S. Muthu (ed.), *Carbon Footprints*, Environmental Footprints and Eco-design
of Products and Processes, https://doi.org/10.1007/978-981-13-7916-1_2

standard consumption units for the period 2006–2013. The databases used are the World Input–Output Database and Spanish Household Budget Survey.

**Keywords** Carbon footprint · Gini index · Palma ratio · Inequality · Consumption patterns · Teleconnection

# 1 Introduction

Palma's work (2011, 2014) shown that in most countries exist extreme inequalities in income distribution. Extreme inequality occurs because the highest income population appropriates a significant part of the income previously received by the lower group of income but not at the expense of the middle group of income, which is relatively stable. This fact makes the gap between rich and poor wider than ever. In this chapter, we propose a measure that allows us to identify the impact that inequality in income distribution has on carbon footprint of households' consumption. The use of a multi-regional input–output model allows us to obtain the environmental footprint associated with households' consumption according to their different levels of income, from the highest income levels to those who are below poverty line and at risk of social exclusion.

Recent literature evaluates the possibility of reducing the impact of household consumption on the environment through the identification and strength of more sustainable consumption patterns using the input–output methodology. These studies show that per capita carbon footprint and other environmental impacts depend not only on income level but also on other non-income factors (such as geography, energy system, production methods, waste management, household size, diet, and lifestyles), so that the effects of increasing income varies considerably between regions (Fleurbaey et al. 2014). Kerkhof et al. (2009) identify some determinants of national households' $CO_2$ emissions and their distribution across income groups in four countries (the Netherlands, UK, Sweden, and Norway). Wiedenhofer et al. (2017) use the carbon-footprint-Gini coefficient to evaluate carbon footprints for Chinese households and find that carbon footprints are unequally distributed among the rich and poor. Hubacek et al. (2014) investigate the potential consequences for climate targets of achieving poverty eradication through a redistribution of income assuming that everybody in the planet has a still modest expenditure level of at least $2.97 power purchasing parity (PPP). Other authors evaluate the impact in household footprints of changes toward a healthier diet (Behrens et al. 2017; Cazcarro et al. 2012), a local sources of food (Weber and Matthews 2008), a local and seasonal consumption (Tobarra et al. 2018), or an aging population (Shigetomi et al. 2014).

Regarding the effect of growth and distribution of income on carbon footprint of Spanish households, we find different studies. Roca and Serrano (2007) analyze the relationship between income growth and nine atmospheric pollutants in Spain taking into account different income levels. Duarte et al. (2010) discuss pollution caused by the Spanish economy and households considering the effects of income inequality across spending levels and establishing a link between the income level, different consumption patterns, consumption propensity, and $CO_2$ emissions. Duarte et al. (2012) evaluate the impact on $CO_2$ emissions of Spanish households' depending on social status (social aspect), level of income (economic dimension), place of residence (rural/urban), and population density (the demographic aspect). López et al. (2016) assess the impact of the Great Recession started in 2008 on the carbon footprint of Spanish households and López et al. (2017) evaluate the effect of income redistribution in the Spanish households' material footprint.

Other authors, using econometric estimations in line of the seminal paper of Boyce (1994), find that more equitable distribution of income involves better environmental quality (only domestic $CO_2$ emissions). This conclusion has been obtained in the short-run and for a set of countries (Ravallion et al. 2000); in the short-run and long-run in the USA (Baek and Gweisah 2013); and for much of Chinese regions (Zhang and Zhao 2014). Similar to this work that takes into account direct and indirect emissions related to household consumption, the study of Lenzen (2003) analyzes the energy use (direct and indirect) derived from consumption of Sydney households' and found a clear correlation between energy use and income.

In this context, the main purpose of this study is to evaluate how social unsustainability, associated with the extreme inequality of income that is transferred through the consumer spending, generates greater environmental unsustainability through the ratio assessment of the households' footprint (more 3.000 euros/up to 1.500 euros). Palma's ratio evolution and its application in terms of the footprint are sensitive to change of number of people and of income they receive. To isolate these two factors, we define two measures: absolute Palma footprint ratio versus a relative measure. The empirical application is carried out on the carbon footprint of Spanish households, combining WIOD and Spanish Households Budget Survey databases. The period of study, 2006–2013, allows us to assess further the impact of the financial and economic crisis. The great financial crisis of 2008 and the subsequent economic crisis did not affect equally to all social classes, displaces an important number of households to lower-income groups and decrease the consumption in these households, more significantly than in the higher-income groups. Finally, a regression analysis was conducted to assess how changes in domestic and imported household consumption and changes in consumption inequalities (Gini and Palma) determine changes in carbon footprint considering the different standard consumption units (SCU) for the period 2006–2013.

This chapter is organized as follows: Sect. 2 describes methods and data sources used; Sect. 3 shows the main results and finally Sect. 4 provides conclusions and some policy recommendations.

# 2 Methodology

In this chapter, we use a multi-regional input–output (MRIO) model to calculate the carbon footprint of households in the Spanish economy.

There are different methods to calculate the carbon footprint. First of all, it should be noted that emissions associated with any activity can be classified as direct or indirect emissions. Direct emissions refer to the emissions generated by burning different types of fuel as a direct consequence of the production process (scope 1) or economic activity. Indirect emissions are a consequence of the activity or production process, but they are not controlled directly by the process, such as the electricity consumption (scope 2), for example, or the remaining indirect emissions associated with the purchase of inputs (scope 3).

Therefore, the calculation of the carbon footprint can be methodologically focused in two directions: the bottom-up approach and the top-down approach. The bottom-up approach is carried out mainly through the application of life-cycle assessment (LCA) models, which are developed with the aim of being applied at the product level considering information about each step of the production processes triggered by the life cycle of the product with a high level of detail. The top-down approach is mainly put into practice usually by applying environmental extended input–output models and specifically MRIO models. In MRIO models, regions and countries are included with its own technology and trade is split into trade of intermediate inputs, with specific industry destination, and trade of final goods, including the global supply chain and all different rounds of production (Johnson and Noguera 2012; Kanemoto et al. 2012; Trefler and Zhu 2010). In addition, there are methodologies that have tried to integrate both perspectives to combine the advantages of both approaches, the so-called hybrid models.

The decision of choosing one methodology or another will depend on the research question that we are addressing, since all models have advantages and disadvantages. In this sense, in this chapter, we use the MRIO model because it allows us to quantify direct and indirect emissions required to satisfy the consumption of households as a part of final demand of the economy (and, therefore, to incorporate scope 3 to the analysis). As it takes into account trade flows, it allows us to identify the importance of domestic and imported suppliers and, therefore, to address the importance of global production chains.

Hence, the MRIO model allows us to assess adequately the impact of inequality on the household's carbon footprint, evaluating the relevance of the global value chains and analyzing the leakage through international trade.

## 2.1 Calculation of Gini Coefficients and Palma Ratio

Gini coefficient and Palma ratio are two measures of inequality used in terms of consumption. The Gini coefficient is calculated as a proportion of the areas in the Lorenz curve. This coefficient takes values between 0 and 1, where 0 corresponds to perfect equality and 1 corresponds to the perfect inequality (Gini 1921). This measure captures the inequality properly when the distribution of income groups a special weight to the middle class is given. However, this measure does not capture adequately the evolution of inequality if it is caused by a change in income in the tails of the distribution (Palma 2014). On the contrary, the ratio of Palma is a useful measure for evaluating the extreme or inter-ratios inequality.

Formally, we define $C$ as the consumption that is performed by the total households and $P$ refers to the number of households for which information is available. The proportion of consumption accumulated for each population group $j$ is given by the expression $c_j = \sum_0^j C_j/C$, for $j = 0, \ldots, n$, with $c_0 = 0$ and $c_n = 1$, and the proportion of the population accumulated by each group is $p_j = \sum_0^j P_j/P$, for $j = 0, \ldots, n$, with $p_0 = 0$ and $p_n = 1$.

Gini coefficient follows the next expression:

$$G = 1 - \sum_{j=1}^{n}(p_j - p_{j-1})(c_j + c_{j-1}) \tag{1}$$

Meanwhile, Palma ratio is obtained as the quotient of the income of the $D10$ (or $D_j$) and the income accumulated by the first four deciles $D1$–$4$ (or $D_i$). A value of 1 in the Palma ratio applied to income is considered an appropriate distribution between the extremes of the distribution (Kharas 2010).

The absolute Palma ratio on household consumption, instead of the original formulation on their income, is given by the following expression:

$$\text{Palma} = C_{10}/C_{1-4} = \sum_9^{10} C_j / \sum_1^4 C_i \tag{2}$$

This absolute measure is suitable for analyzing an extreme distribution of income, but its evolution over time depends on income level per capita and population distribution among different income levels. For instance, if the weight of the population in the extremes of the distribution increases at the expense of a reduction of the middle class, the Palma ratio shows an evolution that depends on changes of individuals from one social class to another one, rather than changes in the distribution of income/consumption between the tails of the distribution. For instance, if there is an impoverishment of the middle class, the reduction of the ratio of Palma that would indicate a reduction of extreme inequality would not be exact. Therefore, we propose a complementary measure, the so-called relative ratio of Palma that assesses the extreme inequality between individuals rather than for an entire society.

The relative Palma ratio is obtained as the quotient between per capita consumption of the decile 10 ($D_j$) and per capita consumption of the decile 1–4 ($D_i$), so its evolution depends only on the average income of a household/individual ($c_{mj}$) representative of the group $j$ in relation to the average income for a representative household ($c_{mi}$) of the group $i$.

$$r\text{Palma} = c_{m10}/c_{m1-4} = \sum_9^{10} C_j/P_j / \sum_1^4 C_i/P_i \tag{3}$$

The importance of defining a relative measure of inequality is related to the existence of a minimum standard of living that households, and only once reached that level of life would allow sustainable consumption decisions.

## 2.2 Environmental Footprint of the Extreme Inequality

The extreme inequality footprint is obtained from the calculation of the carbon footprint associated with the absolute and relative consumption made by each population group identified in a similar way to others authors about carbon footprint (Arce et al. 2017; Druckman and Jackson 2008; Duarte et al. 2012; Hubacek et al. 2017; López et al. 2016; Wiedenhofer et al. 2017).

For the calculation of the carbon footprint associated with the accumulated consumption for each population group $(C_i)$ and to the average per capita consumption $(c_{mi})$ of each population group is required to decompose vector of domestic household consumption into the diagonalized vector of domestic household consumption with the different $j$ characteristics of the region $r$ $(\widehat{C}_i^{rr})$ and the diagonalized vector of imported goods from region $s$ of the same households $(\widehat{C}_i^{sr})$. The expression that assesses the footprint of the households of the group $j$ and the average per capita consumption of the group $j$ of the region $r$ is given by expressions (4) and (5):

$$HCFC_i^r = \begin{pmatrix} f^r & 0 \\ 0 & f^s \end{pmatrix} \begin{pmatrix} L^{rr} & L^{rs} \\ L^{sr} & L^{ss} \end{pmatrix} \begin{pmatrix} \widehat{C}_i^{rr} & 0 \\ 0 & \widehat{C}_i^{sr} \end{pmatrix} = \begin{pmatrix} \overbrace{f^r L^{rr} \widehat{C}_i^{rr}}^{4.1} & \overbrace{f^r L^{rs} \widehat{C}_i^{sr}}^{4.3} \\ \underbrace{f^s L^{sr} \widehat{C}_i^{rr}}_{4.2} & \underbrace{f^s L^{ss} \widehat{C}_i^{sr}}_{4.4} \end{pmatrix} \quad (4)$$

$$HCFC_i^r = \begin{pmatrix} f^r & 0 \\ 0 & f^s \end{pmatrix} \begin{pmatrix} L^{rr} & L^{rs} \\ L^{sr} & L^{ss} \end{pmatrix} \begin{pmatrix} \widehat{c}_i^{rr} & 0 \\ 0 & \widehat{c}_i^{sr} \end{pmatrix} = \begin{pmatrix} \overbrace{f^r L^{rr} \widehat{c}_i^{rr}}^{5.1} & \overbrace{f^r L^{rs} \widehat{c}_i^{sr}}^{5.3} \\ \underbrace{f^s L^{sr} \widehat{c}_i^{rr}}_{5.2} & \underbrace{f^s L^{ss} \widehat{c}_i^{sr}}_{5.4} \end{pmatrix} \quad (5)$$

Therefore, it is important to differentiate between domestic and imported consumption to assess the emissions leakage and track global commodity and value chains via international trade flows (Hubacek et al. 2014; Shigetomi et al. 2014; Yu et al. 2013). Expression (4.1) in (4) shows domestic emissions embodied in production of goods in region $r$ projected to provide final demand of region r. Term (4.2) refers to imported emissions embodied in production of goods in region $r$, projected to provide final demand of the same region $r$. The sum of (4.1) and (4.2) shows household carbon footprint of region $r$ linked to products supplied by industries of region $r$. On the other hand, expression (4.3) shows domestic emissions related with imported goods by household $i$ in region $r$ from other regions $s$. Expression (4.4) refers to imported emissions embodied in imported goods by household $i$ in region $r$ from other regions $s$. Therefore, the addition of (4.3) and (4.4) shows household carbon footprint of region $r$ associated with products imported from region $s$. The interpretation of components of expression (5) is similar. The main difference is that expression (5) informs us about the footprint per household and (4) about the footprint of the set of households.

The proposed formulation could be easily extended to identify the responsibility on inequality of domestic and imported carbon footprint and its evolution over time. The expressions that represent the absolute and relative extreme inequality footprint are, respectively:

$$\text{ExInFoC}_j = \sum_9^{10} \text{HCFC}_j / \sum_1^4 \text{HCFC}_i \tag{6}$$

$$r\text{ExInFoC}_j = \sum_9^{10} \text{HCFc}_{mj} / \sum_1^4 \text{HCFc}_{mi} \tag{7}$$

In a similar way to the Palma ratio, higher values of footprint measures of extreme inequality show that the inequality in consumption inequality is translated in terms of carbon footprint. The difference between absolute and relative measure allows us to isolate changes over time in the measure caused by changes in the extreme inequality footprint, differentiating changes in income from changes in the population belonging to each group. A simple way to isolate the degree to which extreme inequality in consumption moves toward extreme inequality in the carbon footprint would be the quotient between expressions (2) and (6) and the quotient between expressions (3) and (7). If values are greater than 1, it would indicate that the degree of inequality in consumption inequality is greater than in the footprint, as consumption patterns of households with higher-income have a relatively cleaner trend than lower-income consumers.

## 2.3 Regression Analysis

Assessing the environmental sustainability of consumption patterns of different income groups involves analyzing how changes in consumption impact on carbon footprint. To this aim, a regression analysis was conducted differentiating changes in domestic and imported and considering different SCU for the period 2006–2013. The basic equation estimated for household carbon footprint (HCF) is:

$$\text{HCF}_{it} = a + bC_{it} + u_{it} \tag{8}$$

distinguishing between domestic household carbon footprint, associated with domestic consumption $(C_d)$, and imported household carbon footprint, related with imported consumption. To complete the analysis, a dummy variable was introduced to control the effects of time $(T)$, as well as the SCU variable to observe the effects of household size.

Using the estimated equations previously identified, the elasticities are calculated as $\varepsilon^b = \frac{\partial HCF}{\partial C} \frac{\overline{C}}{\overline{HCF}}$; where the term $\frac{\partial HCF}{\partial C}$ is the parameter value $b$ and the variables $\overline{C}$ and $\overline{HCF}$ refers to the mean of consumption and carbon footprint for the whole period.

## 2.4  Data Description

The period analyzed is 2006–2013. We combine information from different statistical sources. Regarding input–output data and environmental information, we used the World Input–Output Database (WIOD) (Timmer et al. 2015) for the period 2006–2011, taking into account the original sectoral aggregation of 35 sectors and 41 regions. For the construction of the final demand vectors, we used the household budget survey (HBS) from the Spanish Statistical Office (INE 2014b).

The surveys used are rich in terms of the variables considered in our analysis and provide information on the household consumption by level of regular net monthly household income, specifically net average monthly household's income. The survey itself provides the information classified in eight levels of income as follows: up to 499 euros, from 500 to 999 euros, from 1000 to 1499 euros, from 1500 to 1999 euros, from 2000 to 2499 euros, from 2500 to 2999 euros, from 3000 to 4999 euros, and 5000 euros or more. The variable chosen in this study is the spending per SCU, although other some magnitudes are provided in terms per capita and per household. Indirect taxes on consumption have been removed using a weighted value-added tax (VAT) rate per year, which has been taken from the Spanish Tax Agency. All the information has been deflated using the consumer price index (CPI) published by the Spanish Statistical Office (INE 2014a) for each year.

Environmental information was obtained from the environmental accounts of WIOD, which provides a set of environmental accounts with detailed information about emissions for country and sector of three greenhouse gases (GHG): $CO_2$, $CH_4$, and $N_2O$. In our analysis, we considered the aggregation of these three GHG measured in $CO_2$ equivalent units, using IPCC global warming potentials; we assume 25 kg/$CO_2$ equivalent per $CH_4$ kg, and 298 kg/$CO_2$ equivalent per $N_2O$ Kg.

## 3  Empirical Results

The rhythm of growth reached by the Spanish economy since the last years of the twentieth century suddenly stopped in 2008 because of the burst of the financial and economic crisis. International crisis—linked to excessive borrowing by Spanish families for house purchases, the inability to deed of assignment in payment, and the need to intervene and rescue entities—lead to stagnation of credit. This fact together with the subsequent economic crisis that ultimately destroys four million jobs from

**Table 1** Gini and Palma indexes of consumption by standard consumption units (SCU), household and per capita, Spain 2006, 2009 and 2013

|  | 2006 | 2009 | 2013 |
|---|---|---|---|
| Gini SCU total (domestic + imported) | 0.144 | 0.138 | 0.150 |
| Gini SCU (domestic) | 0.139 | 0.135 | 0.147 |
| Gini SCU (imported) | 0.194 | 0.175 | 0.196 |
| Gini total (per capita) | 0.107 | 0.104 | 0.120 |
| Gini total (per household) | 0.232 | 0.217 | 0.215 |
| Palma total SCU (domestic + imported) | 0.746 | 1.036 | 0.727 |
| Palma SCU (domestic) | 0.732 | 1.025 | 0.718 |
| Palma SCU (imported) | 0.921 | 1.210 | 0.875 |

*Source* Own elaboration

2006 to 2013 and reduces wages leads to an important decrease of household consumption until 2013. The crisis was an obstacle to total consumption of Spanish families reaching 0.49 billion of euros in 2006. This consumption remains moderate growth of 7% between 2006 and 2009, but the subsequent decline causes that total consumption increased only 2% in 2009 and 2013 (López et al. 2016).

The consumption inequality measured by the Gini index is lower than the inequality associated with the distribution of income. In 2006, Gini index value was 0.107 in per capita terms, 0.143 in SCU terms, and 0.231 in household terms (Table 1). These results are also reflected in the Lorenz curve (Fig. 1). Inequality in consumption distribution also grows during crisis in per capita (0.120) and in SCU (1.150) terms; households, however, can compensate this growth reducing it to 0.214. In other words, crisis-led increase in personal and SCU inequality is partially balanced out by household spending, which is more leveled. Palma ratio is calculated as total consumption made by households earning up to 1499 euros per month between consumption made by those earning above 3000 euros per month for domestic consumption. The extreme measure of inequality is 0.72 for domestic and 0.88 for imported consumption, both cases convey inequalities below 1 and underneath data referred to income (Table 1). Its evolution is the opposite that shown by Gini index, and the variability observed in Palma ratio is bigger; it increases until 2009 and then falls up to 2013 because of the recession. The crisis displaced population of the middle class to lower-middle class and low-income groups, however, the high-income population is maintained and, as a result, the ratio of Palma is reduced.

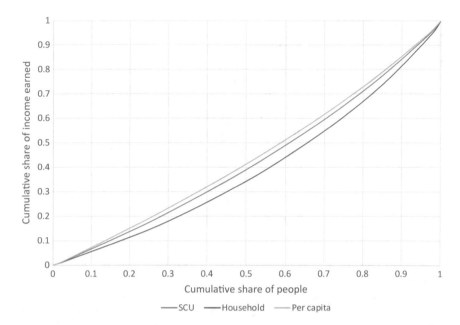

**Fig. 1** Lorenz curve by standard consumption units (SCU), household and per capita, Spain 2013. *Source* Own elaboration

## 3.1 Relative Measure of the Extreme Inequality in the Carbon Footprint

The importance of calculating the carbon footprint by income levels is that it allows us to evaluate the environmental responsibility of different actors. For SCU, 15% of the population with higher-income can explain 33% of the domestic carbon footprint (30% imported) of Spanish households, while 45% of lower-income explains only 23% the domestic carbon footprint (25% imported) in 2013 (Fig. 2). For the whole world economy, Hubacek et al. (2017) found that the top 10% global income earners are responsible for 36% of the current carbon footprint of households, being lower the impact for the Spanish economy, as the inequality in the distribution of income is lower than in many developing countries.

Inequality measures from the carbon footprint maintain a degree of inequality similar to those calculated on the consumption, and these are consistent with the elasticities around 1 previously presented. In 2013, the calculation of the Gini index for the carbon footprint of the SCU was 0.147 and, for consumption, 0.150 (Table 2). In a similar way, the inequality shown by the Gini index of domestic carbon footprint of SCU is smaller than the imported, with a value of 0.139. Regarding the absolute extreme inequality measure, consumption of the total households earning up to 1499 euros per month divided between those who earn more than 3000 euros (P85/P1–45) shows a result of 0.71 for domestic carbon footprint and 0.82 for domestic carbon

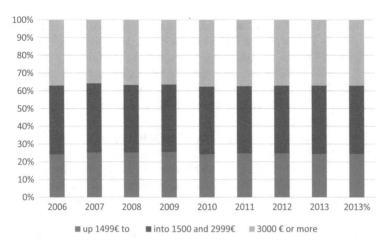

**Fig. 2** Carbon footprint by income groups of standard consumption units, Spain 2006–2013. *Source* Own elaboration

**Table 2** Absolute and relative extreme inequality of carbon footprint by consumer groups, Spain, 2006–2013

|                        | 2006 | 2007 | 2008 | 2009 | 2010 | 2011 | 2012 | 2013 |
|------------------------|------|------|------|------|------|------|------|------|
| Total (domestic)       | 0.74 | 0.97 | 1.1  | 1.03 | 0.85 | 0.91 | 0.78 | 0.71 |
| Total (imported)       | 0.85 | 1.09 | 1.24 | 1.15 | 0.94 | 1.05 | 0.89 | 0.82 |
| SCU (global)           | 1.53 | 1.42 | 1.45 | 1.42 | 1.55 | 1.51 | 1.5  | 1.51 |
| SCU (domestic)         | 1.45 | 1.37 | 1.4  | 1.38 | 1.49 | 1.46 | 1.45 | 1.47 |
| SCU (imported)         | 1.84 | 1.67 | 1.7  | 1.62 | 1.81 | 1.78 | 1.73 | 1.76 |
| Households (global)    | 2.37 | 2.23 | 2.26 | 2.08 | 2.21 | 2.15 | 2.08 | 2.06 |
| Households (domestic)  | 2.26 | 2.15 | 2.17 | 2.03 | 2.13 | 2.07 | 2.02 | 1.99 |
| Households (imported)  | 2.81 | 2.58 | 2.61 | 2.35 | 2.57 | 2.52 | 2.4  | 2.38 |
| Per capita (global)    | 1.26 | 1.16 | 1.19 | 1.22 | 1.33 | 1.3  | 1.3  | 1.32 |
| Per capita (domestic)  | 1.45 | 1.37 | 1.4  | 1.38 | 1.49 | 1.46 | 1.45 | 1.47 |
| Per capita (imported)  | 1.54 | 1.38 | 1.4  | 1.39 | 1.56 | 1.54 | 1.5  | 1.54 |

*Source* Own elaboration

footprint and therefore less than 1. That is, the inequality in consumption is transferred very similarly to the domestic carbon footprint, but, however, is amplified regarding imported carbon footprint.

Information from the household budget survey allows us to transfer this inequality of economic agents to different consumption agents when dividing by the number of individuals, households, and SCU. The main conclusion is that relative extreme inequality is higher than the absolute inequality measure. In 2013, the ratio of Palma by consumption agents for the domestic and imported carbon footprint per capita is 1.32, 1.51 for the SCU and 2.06 for households (Table 2). Inequality in consumption and carbon footprint grows as households focus on people who have a higher-income and, however, is equalized in the SCU as their size determines their spending the same way as the Gini index consumption and carbon footprint. That is, each of the 15 percentile SCU is responsible for a carbon footprint that is 1.51 times higher than the P1–45 SCU. Similarly, SCU inequality is higher for calculations related to the carbon footprint of imported consumption (1.76) to address the carbon footprint of domestic consumption (1.46).

Time trends show how inequality in SCU through the extreme inequality measure is reduced until 2009 (1.42) and then with the deepening crisis in 2013 returns to values greater inequality 2006 (1.53). This result is the opposite the one found in the measurement of absolute extreme inequality measure and the reason is that in the growth phase and up to 2009 the percentage of households with lower-incomes or to 1499 euros per month is reduced (going from 42 to 37%) and then with the deepening crisis increases the percentage of lower-income families (45%).

## 3.2 Inequality and Carbon Footprint: A Regression Analysis

Assessing the environmental sustainability of consumption patterns of different income groups involves analyzing how changes in the consumption have an impact on carbon footprint. We find a value for the elasticity of 0.934 between the domestic household consumption and its carbon footprint and an elasticity of 0.967 for imported consumption and its carbon footprint. These results indicate that consumption increases generate less than proportional increases in carbon footprint (Table 3). The incorporation of time using a dummy variable does not greatly affect to the results, just as occurs with the inclusion of households' size—as occurs in Jones and Kammen (2011), which change the elasticity of domestic consumption to 0.9339 and 0.9436, respectively. Considering both variables simultaneously, the elasticity raise to 0.9424 (for carbon footprint associated with households' imported consumption figures are 0.9316, 1.003 and 0.9294). These results are higher than those found by Duarte et al. (2012) for domestic carbon footprint of the Spanish economy in 1999, around 0.5; also for domestic and imported US carbon footprint in 2004 estimated by Weber and Matthews (2008), 0.4 and 0.6, respectively, or for Lenzen et al. (2006) who examine total primary energy and obtain an elasticity of 0.78, indicating that the energy intensity diminishes toward higher expenditure. These lower elasticities are explained using HBS micro-data, where the effects of household size on the carbon footprint are more perceived.

In terms of teleconnection, the elasticity of carbon footprint associated with consumption of domestic goods is higher than the elasticity of the household carbon footprint of imported goods, as is found in Weber and Matthews (2008). That is, when an increase of the level of consumption by different income groups is given, a change in the consumption patterns occurs, where domestic consumption is more sustainable than imported one in terms of emissions.

**Table 3** Regression results for household carbon footprint (HCF) groups, Spain 2006–2013

| | Domestic HCF | | | | Imported HCF | | | |
|---|---|---|---|---|---|---|---|---|
| Intercept | 0.2223** | 0.421975 | 0.052261 | 0.232598 | 0.0391· | 0.4166* | 0.0177 | −0.0029 |
| $C_d$ (ε) | 0.1911*** (0.934) | 0.1931*** (0.9436) | 0.1911*** (0.9338) | 0.1929*** (0.9424) | | | | |
| $C_m$ (ε) | | | | | 0.9217*** (0.9678) | 0.9555*** (1.003) | 0.8872*** (0.9316) | 0.8851*** (0.9294) |
| T06 | | | 0.5797*** | 0.5790*** | | | 0.1963*** | 0.1973*** |
| T07 | | | 0.3575*** | 0.3554*** | | | 0.1668*** | 0.1677*** |
| T08 | | | 0.001876 | −0.0009 | | | 0.0661*** | 0.0668*** |
| T09 | | | 0.1810*** | 0.1804*** | | | −0.0001043 | −0.0001211 |
| T10 | | | 0.1391** | 0.1382** | | | 0.0398** | 0.0400** |
| T11 | | | 0.0759· | 0.0754 | | | 0.0376** | 0.0378** |
| T12 | | | 0.0291 | 0.0287· | | | 0.0175 | 0.0176 |
| SCU | | −0.1519 | | −0.1366 | | −0.27559* | | 0.015 |
| Residual standard error | 0.2086 on 62 DF | 0.21 on 61 DF | 0.08297 on 55 DF | 0.08302 on 54 DF | 0.0723 on 62 DF | 0.06981 on 61 DF | 0.02425 on 55 DF | 0.02445 on 54 DF |
| Multiple R-squared | 0.9644 | 0.9645 | 0.995 | 0.9951 | 0.9828 | 0.9843 | 0.9983 | 0.9983 |
| Adjusted R-squared | 0.9638 | 0.9633 | 0.9943 | 0.9943 | 0.9826 | 0.9838 | 0.998 | 0.998 |
| F-statistic | 1680 on 1 and 62 DF | 828.9 on 2 and 61 DF | 1370 on 8 and 55 DF | 1216 on 9 and 54 DF | 3552 on 1 and 62 DF | 1908 on 2 and 61 DF | 4009 on 8 and 55 DF | 3506 on 9 and 54 DF |
| p-value | <2.2e−16 | <2.2e−16 | <2.2e−16 | <2.2e−16 | <2.2e−16 | <2.2e−16 | <2.2e−16 | <2.2e−16 |

*Note* *Signif. codes: 0 '\*\*\*' 0.001 '\*\*' 0.01 '\*' 0.05 '.' 0.1 ' ' 1
*Source* Own elaboration

**Table 4** Regression results for household carbon footprint (HCF) related with domestic consumption, Spain 2006–2013

| HCF related with domestic consumption | | | | | | |
|---|---|---|---|---|---|---|
| Intercept | 1.8037* | 0.9352· | 1.5794* | 0.9068· | 1.7977* | 0.6623 |
| $C_d$ | 0.1909*** | 0.1908*** | 0.1908*** | 0.1908*** | 0.1909*** | 0.19113*** |
| PalmaD | −1.0732* | −2.7767*** | | | −1.7834 | −7.6307*** |
| PalmaM | | 1.7929*** | | | | 2.4558** |
| GiniD | | | −9.7456* | −50.4699*** | 7.5587 | 54.3606* |
| GiniM | | | | 34.0473*** | | −7.48 |
| Residual standard error | 0.202 on 61 DF | 0.1407 on 60 DF | 0.2037 on 61 DF | 0.15 on 60 DF | 0.2033 on 60 DF | 0.1261 on 58 DF |
| Multiple R-squared | 0.9672 | 0.9843 | 0.9666 | 0.9822 | 0.9673 | 0.9878 |
| Adjusted R-squared | 0.9661 | 0.9835 | 0.9655 | 0.9813 | 0.9656 | 0.9868 |
| F-statistic | 898.4 on 2 and 61 DF | 1256 on 3 and 60 DF | 883 on 2 and 61 DF | 1103 on 3 and 60 DF | 591.3 on 3 and 60 DF | 942.3 on 5 and 58 DF |
| p-value | <2.2e−16 | <2.2e−16 | <2.2e−16 | <2.2e−17 | <2.2e−16 | <2.2e−16 |

*Note* *Signif. codes: 0 '***' 0.001 '**' 0.01 '*' 0.05 '.' 0.1 ' ' 1
*Source* Own elaboration

The elasticities found for marginal increases reach values of 0.99 for domestic carbon footprint and 1.01 for carbon footprint related with households' imported consumption. That is, after eliminating the standard consumption associated with the level of income of the poorest households (the first level of income), marginal increases in consumption lead to increases in carbon footprint that are more polluting on average than the first group of consumption, especially the imported household carbon footprint, because of having the elasticity higher than 1.

An additional contribution is the consideration of different inequality measures within the econometric analysis, such as Gini index and Palma index, differentiating between domestic and imported (Tables 4 and 5). The inclusion of these coefficients allows seeing how inequality (in terms of consumption) is affecting carbon footprint associated with domestic and imported consumption and how the increase in inequality that has occurred in Spain since 2009, because of the economic crisis, impact on consumption patterns. A positive sign of the inequality variable (Gini or Palma index) would indicate that an increase in inequality increases carbon footprint and, by contrast, a negative sign indicates that an increase in inequality reduces carbon footprint.

The first place it can be observed as consumption remains positive and significant in all cases when we introduce measures of inequality (Tables 4 and 5). Carbon footprint of domestic goods grow when inequality in consumption of domestic goods

**Table 5** Regression results for household carbon footprint (HCF) related with imported consumption, Spain 2006–2013

| HCF related with imported consumption | | | | | | |
|---|---|---|---|---|---|---|
| Intercept | −0.5855*** | 0.5217*** | −0.1161 | 0.580461*** | −0.04782 | 0.641216*** |
| $C_m$ | 0.9152*** | 0.8900*** | 0.9219*** | 0.890063*** | 0.90259*** | 0.888783*** |
| PalmaM | 0.3365*** | 0.7311*** | | | 0.88614*** | 0.215806 |
| PalmaD | | −1.2368*** | | | | −1.424456*** |
| GiniM | | | 0.8347 | 14.473895*** | −8.37592*** | 11.644463** |
| GiniD | | | | −22.981397*** | | −7.4612 |
| Residual standard error | 0.06536 on 61 DF | 0.03564 on 60 DF | 0.07262 on 61 DF | 0.0356 on 60 DF | 0.05502 on 60 DF | 0.03099 on 58 DF |
| Multiple R-squared | 0.9862 | 0.996 | 0.983 | 0.996 | 0.9904 | 0.9971 |
| Adjusted R-squared | 0.9858 | 0.9958 | 0.9824 | 0.9958 | 0.9899 | 0.9968 |
| F-statistic | 2181 on 2 and 61 DF | 4939 on 3 and 60 DF | 1761 on 2 and 61 DF | 4951 on 3 and 60 DF | 2060 on 3 and 60 DF | 3923 on 5 and 58 DF |
| p-value | <2.2e−16 | <2.2e−16 | <2.2e−16 | <2.2e−16 | <2.2e−16 | <2.2e−16 |

*Note* *Signif. codes: 0 '***' 0.001 '**' 0.01 '*' 0.05 '.' 0.1 ' ' 1
*Source* Own elaboration

increases. An increase in domestic inequality would reduce the domestic household carbon footprint. Domestic measures (PalmaD and GiniD) present a negative sign and are significant in both cases (Table 4) because the SCU with higher-income increases the consumption of domestic goods and services at lower rates than low-income families (Table 4). This fact would imply that the search for growth that seeks to reduce income inequality in Spain, to improve the social sustainability, would be unsustainable in terms of carbon footprint. However, the introduction in the domestic household carbon footprint estimation of imported Palma and Gini indexes must be interpreted as "cross-inequality measures", because they would be reflecting the substitution effect between domestic and imported consumption. These measures present a positive sign, showing that an increase in inequality in consumption of imported goods and services generates an increase in carbon footprint linked to domestic consumption. The reason is that high-income households can replace its domestic consumption by imported one and as households have more income and choose to consume more imported goods, they consume less domestic goods and services, leading to a reduction in carbon footprint related with domestic consumption. Therefore, these results show how an increase in inequality in the consumption of the SCU in the Spanish economy, between 2006 and 2013, causes a change in consumption patterns, substituting domestic consumption by imported consumption.

Table 5 shows household carbon footprint associated with the consumption of imported goods and services, and these estimations confirm the argument put forward about changes in consumption patterns. Household carbon footprint of imported goods decreases when inequality in consumption of imported goods increases.

Imported Palma and Gini indexes (PalmaM and GiniM) show a positive sign in both estimations. The less inequality in consumption of imported goods means a lower carbon footprint linked to consumption of imported goods and services, because the high-income households are those that maintain a higher propensity to import. These results are similar to that found by authors such as Ravallion et al. (2000), Baek and Gweisah (2013) and Zhang and Zhao (2014) that find that more equitable distribution of income in the results in better environmental quality (domestic $CO_2$ emissions). However, when the "cross-inequality measures" are introduced, in this case are the domestic Palma and Gini indexes (PalmaD and GiniD), the sign obtained is negative: the lower inequality in consumption, the higher imported carbon footprint. This is explained by the ability of households with higher-income to substitute domestic consumption by imported consumption. During the economic crisis, they reduce the consumption of superfluous goods, which are mainly imported, and maintain the basic consumption (food, energy, and housing) that are more local (although these remain an indirect leakage through imports of intermediate goods).

In both cases, household carbon footprint linked to domestic and imported consumption, the inclusion in the estimation of variables of inequality maintains the sign and significance (Table 4 and 5). In light of the results, Palma index seems to be more appropriate for assessing the inequality impact since in all estimates maintains the sign and is significant. It is just not significant when Gini index is introduced, which indicate that both indexes are measuring the same thing, so that offset each other.

Hence, it is important to highlight the different weights that the domestic and imported consumption have at different income levels. As previous sections already put in evidence, the lower-income level, the most important is domestic consumption over total consumption of the income group because they increase its imported consumption more importantly than their domestic consumption. In the period focus of the study, 2006–2013, the inequality first decreases between 2006 and 2009, and then it started growing until 2013. The most significant fact during the crisis is that when inequality grows households do not reduce in the same proportion its domestic and imported consumption but reduced in a larger share the imported consumption (rising from 30 to 20% for the group 8); moreover, lower-income households that increase their weight in the consumption keep more important domestic consumption share than households with higher-income.

Finally, due to the different impacts of inequality regarding household carbon footprint associated with domestic and imported consumption found in the previous analysis, we performed an estimation that evaluates the effect of total household carbon footprint (domestic and imported) in relation to total consumption and two measures of overall inequality (Table 6). Again, in this case, Gini coefficient is significant and negative and Palma ratio is only significant and negative when it is introduced together with Gini coefficient. Between 2006 and 2013, as the inequality is growing, carbon footprint of Spanish households is reduced and thus, more sustainable growth in social terms, or less unequal, would result in more unsustainable world in terms of carbon footprint. The higher weight that the carbon footprint associated with domestic consumption (75%) over the imported consumption (the

**Table 6** Net effects of the inequality measures for total household carbon footprint (HCF), Spain 2006–2013

| Total HCF | | | | |
|---|---|---|---|---|
| Intercept | 0.005524 | −0.04078 | 5.7671** | 34.59*** |
| C | 0.350908*** | 0.35087*** | 0.3495*** | 0.3489*** |
| Palma | | 0.05517 | | −7.269*** |
| Gini | | | −40.3535** | −1995*** |
| Residual standard error | 0.6256 on 62 DF | 0.6307 on 61 DF | 0.5942 on 61 DF | 0.432 on 60 DF |
| Multiple R-squared | 0.9246 | 0.9246 | 0.933 | 0.9652 |
| Adjusted R-squared | 0.9234 | 0.9221 | 0.9308 | 0.9635 |
| F-statistic | 760 on 1 and 62 DF | 373.9 on 2 and 61 DF | 425 on 2 and 61 DF | 554.7 on 3 and 60 DF |
| p-value | <2.2e−16 | <2.2e−16 | <2.2e−16 | <2.2e−16 |

*Note* *Signif. codes: 0 '***' 0.001 '**' 0.01 '*' 0.05 '.' 0.1 ' ' 1
*Source* Own elaboration

remaining 25%) is generated as a result that in this net estimation the negative sign is predominant.

# 4 Conclusions

The economic crisis in the Spanish economy started in 2008 generated an increase in relative extreme inequality that was transferred to the carbon footprint of households. Households that spent more than 3000 euros maintain a higher carbon footprint than households spending less than 1500 euros between 2008 and 2013. The crisis, though it affected all households, affected more intensely to lower-income households because they reduce their consumption more than higher-income households. However, both the household consumption group and the number of households in each group changed over time. In the Spanish economy, the crisis also moved many households from the middle class to lower-middle classes who are below the poverty line.

The analysis of the results from the regression it can be concluded that, between 2006 and 2013, as the inequality grew, carbon footprint of Spanish households was reduced, and, therefore, more sustainable growth in social terms (or less unequal) would have resulted in more unsustainable world in terms of carbon footprint. The difficulty of achieving the reduction of inequality and the reduction of the carbon footprint at the same time leads us to suggest measures that encourage the modification of the consumption basket toward a more sustainable consumption pattern.

Therefore, raising awareness and modifying the behavior of the 15% of the population who receives higher-income (and perform the highest consumption) is very important, as their responsibility in terms of carbon footprint reaches to represent more than 30% of the total of the households' carbon footprint. Training, awareness policies, and information campaigns should be developed especially for these groups because they have the level of income that allows them to make sustainable consumption decisions. A reduction of inequality in income distribution would allow to lower-income families make decisions for a sustainable consumption, if they reach the minimum standard of living that allow access sustainable decisions.

# References

Arce G, Zafrilla JE, López LA, Tobarra MÁ (2017) Carbon footprint of human settlements in Spain. In: Álvarez R, Zubelzu S, Martínez R (eds), Carbon footprint and the industrial lifecycle—from urban planning to recycling

Baek J, Gweisah G (2013) Does income inequality harm the environment? Empirical evidence from the United States. Energy Policy 62:1434–1437

Behrens P, Kiefte-de Jong JC, Bosker T, Rodrigues JFD, de Koning A, Tukker A (2017) Environmental impacts of dietary recommendations. Proc Natl Acad Sci 114(51):13412–13417. 10.1073/pnas.1711889114

Boyce JK (1994) Inequality as a cause of environmental degradation. Ecol Econ 11:169–178

Cazcarro I, Duarte R, Sánchez-Chóliz J (2012) Water flows in the spanish economy: agri-food sectors, trade and households diets in an input-output framework. Environ Sci Technol 46:6530–6538

Druckman A, Jackson T (2008) Measuring resource inequalities: the concepts and methodology for an area-based Gini coefficient. Ecol Econ 65:242–252

Duarte R, Mainar A, Sánchez-Chóliz J (2010) The impact of household consumption patterns on emissions in Spain. Energy Econ 32:176–185

Duarte R, Mainar A, Sánchez-Chóliz J (2012) Social groups and $CO_2$ emissions in Spanish households. Energy Policy 44:441–450

Fleurbaey M, Kartha S, Bolwig S, Chee YL, Chen Y, Corbera E, Lecocq F, Lutz W, Muylaert MS, Norgaard RB, Okereke C, Sagar AD (2014) Sustainable development and equity. In: Climate change 2014: mitigation of climate change. contribution of working group III to the fifth assessment report of the intergovernmental panel on climate change. Cambridge University Press, Cambridge, United Kingdom and New York, NY, USA

Gini C (1921) Measurement of inequality of incomes. Econ J 31:124–126

Hubacek K, Baiocchi G, Feng K, Patwardhan A (2017) Poverty eradication in a carbon constrained world. Nat Commun 8:912

Hubacek K, Feng K, Minx JC, Pfister S, Zhou N (2014) Teleconnecting consumption to environmental impacts at multiple spatial scales. J Ind Ecol 18:7–9

INE (2014a) Índice de Precios al Consumo. Retrieved from http://www.ine.es/

INE (2014b) Encuesta de Presupuestos Familiares. Retrieved from http://www.ine.es/

Johnson RC, Noguera G (2012) Accounting for intermediates: production sharing and trade in value added. J Int Econ 86:224–236

Jones CM, Kammen DM (2011) Quantifying carbon footprint reduction opportunities for U.S. households and communities. Environ Sci Technol 45:4088–4095

Kanemoto K, Lenzen M, Peters GP, Moran DD, Geschke A (2012) Frameworks for comparing emissions associated with production, consumption and international trade. Envion Sci Technol 46:172–179

Kerkhof AC, Benders RMJ, Moll HC (2009) Determinants of variation in household $CO_2$ emissions between and within countries. Energy Policy 37:1509–1517

Kharas H (2010) The emerging middle class in developing countries. In: Global development outlook. OECD Development Centre

Lenzen M (2003) Environmentally important paths, linkages and key sectors in the Australian economy. Struct Change Econ Dyn 14:1–34

Lenzen M, Wier M, Cohen C, Hayami H, Pachauri S, Schaeffer R (2006) A comparative multivariate analysis of household energy requirements in Australia, Brazil, Denmark, India and Japan. Energy 31:181–207

López LA, Arce G, Morenate M, Monsalve F (2016) Assessing the inequality of spanish households through the carbon footprint: the 21st century great recession effect. J Ind Ecol 20:571–581

López LA, Arce G, Morenate M, Zafrilla JE (2017) How does income redistribution affect households' material footprint? J Clean Prod 153:515–527

Palma JG (2011) Homogeneous middles versus heterogeneous tails, and the end of the 'Inverted-U': it's all about the share of the rich. Dev Change 42:87–153

Palma JG (2014) Has the income share of the middle and upper-middle been stable around the '50/50 Rule', or has it converged towards that level? the 'Palma Ratio' revisited. Dev Change 45:1416–1448

Ravallion M, Heil M, Jalan J (2000) Carbon emissions and income inequality. Oxford Econ Pap 52:651–669

Roca J, Serrano M (2007) Trade and atmospheric pollution in spain: an input-output approach. In: 2nd Spanish conference on input-output analysis, Zaragoza, Spain, 5–7 Sept 2007

Shigetomi Y, Nansai K, Kagawa S, Tohno S (2014) Changes in the carbon footprint of Japanese households in an aging society. Environ Sci Technol 48:6069–6080

Timmer MP, Dietzenbacher E, Los B, Stehrer R, de Vries GJ (2015) An illustrated user guide to the world input-output database: the case of global automotive production. Rev Int Econ 23:575–605

Tobarra MA, López LA, Cadarso MA, Gómez N, Cazcarro I (2018) Is seasonal households' consumption good for the nexus carbon/water footprint? The Spanish fruits and vegetables case. Environ Sci Technol 52:12066–12077

Trefler D, Zhu SC (2010) The structure of factor content predictions. J Int Econ 82:195–207

Weber CL, Matthews HS (2008) Food-miles and the relative climate impacts of food choices in the United States. Environ Sci Technol 42:3508–3513

Wiedenhofer D, Guan D, Liu Z, Meng J, Zhang N, Wei Y-M (2017) Unequal household carbon footprints in China. Nat Clim Change 7:75

Yu Y, Feng K, Hubacek K (2013) Tele-connecting local consumption to global land use. Glob Environ Change 23:1178–1186

Zhang C, Zhao W (2014) Panel estimation for income inequality and $CO_2$ emissions: a regional analysis in China. Appl Energy 136:382–392

# Software for Calculation of Carbon Footprint for Residential Buildings

Jaime Solís-Guzmán, Cristina Rivero-Camacho, Mónica Tristancho, Alejandro Martínez-Rocamora and Madelyn Marrero

**Abstract** Predicting the environmental impact of buildings from early stages of design becomes essential for improving their sustainability during the entire life cycle. Carbon footprint allows to determine the emissions of greenhouse gases derived from the building process from cradle to grave. In this chapter, the experience of the authors in carbon footprint calculation methodologies is presented through an open-source software tool for estimating the environmental impact of architectural projects from the design phase, as a product from the OERCO2 Erasmus + project. With a newly renovated interface, the internal functioning of this tool is comprehensively explained, presenting and analyzing the results obtained for representative building typologies. This tool aims to allow users to detect opportunities of improvement, environmental and economic, of their projects through modifications such as the selection of different materials or constructive solutions. Subsequently, the authors carry out an analysis of its flaws and forthcoming necessary upgrades to make this software tool more accessible to non-specialized users, thus easing the spread of knowledge on carbon footprint and the environmental impact of buildings.

**Keywords** Carbon footprint · Emissions · Construction · Building · Resources · Consumption

## 1 Introduction

The environmental impact indicators reported in the latest years highlight the construction sector as one of the main energy consumers and $CO_2$ emission generators among the various industrial sectors, with estimations of 30–40% of the total

J. Solís-Guzmán (✉) · C. Rivero-Camacho · M. Tristancho · M. Marrero
Departamento de Construcciones Arquitectónicas II. ETS Ingeniería de Edificación, Universidad de Sevilla, Avenida Reina Mercedes nº 4, 41012 Seville, Spain
e-mail: jaimesolis@us.es

A. Martínez-Rocamora
GACS Research Group, Department of Construction Sciences, Faculty of Architecture, Construction and Design, University of Bío-Bío., Av. Collao, 1202, Concepción, Chile

© Springer Nature Singapore Pte Ltd. 2020 55
S. S. Muthu (ed.), *Carbon Footprints*, Environmental Footprints and Eco-design of Products and Processes, https://doi.org/10.1007/978-981-13-7916-1_3

environmental impact produced by society (European Parliament—Council of the European Union 2018). This concern has forced the emergence of international standards to promote the use of environmental labels of building products (UNE-EN ISO 14020 2002; UNE-EN ISO 14025 2006; UNE-EN 15804 2012; UNE-EN ISO 14021 2017), the development and application of life cycle assessment (LCA) in this sector (UNE-EN ISO 14040 2006; UNE-EN ISO 14044 2006; UNE-EN 15978 2012), and the environmental management of buildings during from a lifecycle perspective (UNE-EN ISO 14001 2015; ISO 15686-5 2017). These aims are not always easy to accomplish due to economic, technical, practical, and cultural barriers that prevent professionals from selecting more environment-friendly materials (Giesekam et al. 2016).

Results obtained from the various methodologies of LCA-based indicators applied to the construction sector have the urgent need to be easily communicated to the non-specialized society. Among all the ecological indicators developed in the latest decades, carbon footprint (CF), an indicator of the greenhouse gas emissions generated by a determined process (Weidema et al. 2008), stands out for its simplicity and relation with the main aims of the Kyoto Protocol (Cagiao et al. 2011), along with its easy application in decision-making and environmental policy (Bare et al. 2000). Although it is not easy to adapt ecological indicators to the construction sector, a considerable amount of proposals can be found in the scientific literature (Geng et al. 2017), whose results do not always match, mainly due to the absence of an exact calculation methodology in the current international standards (Dossche et al. 2017).

Most of the recent studies proposing methodologies to estimate the environmental impact of buildings or applying ecological indicators to case studies of buildings have been collected in several reviews focused on LCA (Buyle et al. 2013), lifecycle energy analysis (Ramesh et al. 2010) and lifecycle carbon footprint (Schwartz et al. 2018), and a variety of indicators (Cabeza et al. 2014; Chau et al. 2015). The methodologies extracted from these studies serve as a basis to develop calculation tools that measure the environmental impact of buildings to allow the appearance of the environmental certification systems and consequently environmental policies based on them.

In Spain, there is a variety of these tools that somehow include the calculation of carbon footprint of buildings. The most important tools are LEED and BREEAM, whose use has spread in our country thanks to national organisms such as the Spain Green Building Council (SpainGBC 2015) and BREEAM Spain (BREEAM 2017). These tools include among the various aspects evaluated to obtain a final score the $CO_2$ emissions from the manufacturing of building materials and the operational energy; however, that final score does not reflect these $CO_2$ emissions, thus failing in communicating every result separately for the sake of a better understanding and the subsequent analysis of possible improvements.

But other alternatives have arisen from several research projects in the latest decade in Spain. For example, SpainGBC presented the VERDE tools (SpainGBC 2013), a set of environmental impact assessment tools for design assistance (HADES), new buildings (VERDE NE), rehabilitation (VERDE RH), and urban development (VERDE DU). In this set of tools, carbon footprint gets the highest score percentage, thus prevailing over other environmental impact sources. ECOMETRO is a Web-based open-source tool to measure the environmental impact of a building (Asociación Ecómetro 2017). It is similar to an EPD, but applied to entire buildings.

Other certification tools focus on the energy demand during the operational phase of the building's life cycle. In Spain, the recognized tools for energy certification of buildings, open and free accessed are CE3, CE3X, CERMA (Spain METDA 2017), the Unified Tool LIDER-CALENER (Spain Ministry of Development 2015), developed by Spanish associations and universities, and private is CYPETHERM (Cype Ingenieros SA 2018), created by CYPE Ingenieros. The estimates of $CO_2$ emissions obtained from these tools do not take into account the embodied $CO_2$ of the construction materials consumed in the building.

Highly specialized platforms such as BEDEC cost database, SOFIAS tool, or e2CO2Cero allow detailed calculation of $CO_2$ emissions based on the project's bill of quantities. BEDEC was developed by the Institute of Construction Technology of Catalonia (ITeC) and uses environmental data of construction materials from the Ecoinvent LCA database (Ecoinvent Centre 2013), well-known for being one of the most comprehensive databases at European level (Martínez-Rocamora et al. 2016a) and for its integration with the SimaPro LCA software (PRé Sustainability 2016). SOFIAS tool, on the other hand, uses data from the OpenDAP database (SOFIAS Project 2017). As an intermediate solution, there is e2CO2Cero, by the Basque Government, a software that allows to estimate the embodied energy and carbon footprint of a building according to the materials consumed and the construction processes used for that phase of the life cycle (e2CO2cero 2014). This too has the peculiarity of offering two different versions: complete and simplified, the former requiring to introduce the project's bill of quantities, which is considered the appropriate way to reach the general public and create social awareness.

Most of these tools require specialized knowledge on construction and energy sources, or even environmental knowledge, and as a consequence, non-specialized users are incapable of using them in order to get a slight idea of how sustainable is their own house or building. This kind of information would be useful to get orientation about how to improve the environmental behavior of their houses (Marrero et al. 2018). But an intermediate public must also be regarded, since non-specialized users can also be defined as those that, even having technical knowledge (architecture or construction students, professionals from the construction sector, etc.), have not been trained to assess the environmental impacts generated in construction processes.

This is where open educational resources (OERs hereafter) fit as an adequate solution for this intermediate step in the education on environmental impact assessment methodologies, providing (as a public good) free open learning tools to spread the knowledge generated in this research field (Tovar et al. 2013; Zancanaro et al. 2015). From the publication of the OpenCourseWare (OCW) by the MIT, numerous OERs have been developed in order to adapt education methodologies by including the use of new available technologies such as mobile phones, tablets, video, and test platforms as part of the learning process (Tovar and Piedra 2014). Thus, a considerable number of OER repositories have emerged in the Internet covering knowledge on all kinds of subject and grades, from pre-school to higher education (Di Blas et al. 2014).

Challenges and needs for the implementation, sharing, and use of OERs have been thoroughly analyzed (Chen 2010), establishing proposals for IT frameworks (Khanna and Basac 2013) and strategies to facilitate the effective use of this technology by improving its quality, reusability, and integration in the teaching process. Regarding quality, (Clements and Pawlowski 2012) carried out an interesting study on the perception of quality that teachers and students considered important for OERs and determined that trust in the developing organism and easiness of adapting the contents to their own purposes were two of the most important factors to choose a specific OER.

Some authors have pointed out the usefulness of recording lectures for implementing flipped classroom methodologies, reaching students with disabilities, or spreading knowledge through the OCW initiative (Llamas-Nistal and Mikic-Fonte 2014). Moreover, OERs have other additional advantages over traditional learning, such as the possibility to have a massive number of students taking a course at the same time (Tuomi 2013), or to tackle financial disadvantages of students by replacing textbooks with Web-based open courses (McGreal et al. 2012; Nipa and Kermanshachi 2018). Through a quantitative and qualitative analysis, Nipa and Kermanshachi concluded that financially disadvantaged students had a more positive perception of the OER course materials, and students using OER materials received higher grades than those using traditional textbook-based materials.

In this chapter, the experience of the authors in carbon footprint calculation methodologies is presented through an open-source software tool for estimating the environmental impact of architectural projects from the design phase, as a product from the OERCO2 Erasmus + project. With a newly renovated interface, the internal functioning of this tool is comprehensively explained, presenting and analyzing the results obtained for representative building typologies.

First, the OERCO2 project experience and activities are presented which include an analysis of the environmental product declarations available in the participating countries, the analysis of $CO_2$ calculation methodologies, surveys in each country in order to determine the actual knowledge on $CO_2$ assessment and the dissemination activities. Secondly, the $CO_2$ assessment tool is explained in detail and different construction projects are studied in order to see its teaching potential.

## 2  Materials and Methods

### 2.1  OERCO2 Project

The OERCO2 project (OERCO2 2018) was funded by the Erasmus + program in 2016 within the scope of Strategic Associations in the Higher Education sector (KA203) (SEPIE 2017). The University of Seville (Spain) led the project, and the partners were the Technological Center of the Marble (Murcia, Spain), CERTIMAC

(Faenza, Italy), Green Building Council (Bucharest, Romania), CTCV (Coimbra, Portugal), and the University of Transylvania (Brasov, Romania).

The main aims of the project included:

- Studying the methodology for the calculation of $CO_2$ emissions of the construction process and throughout the life cycle of materials at European level;
- Establishing a common European curriculum in this area, thus increasing awareness on climate change and providing information on the emissions generated by each element;
- Developing an open educational resource (OER) to spread knowledge on $CO_2$ emissions in construction processes;
- Launching an online tool accessible to all agents in the construction sector (students, professionals, etc.) at European level which does not require previous knowledge on environmental impact assessment for its use.

In order to achieve these goals, four milestones had been established in the project, targeting the participating countries of the OERCO2 project: (1) a state-of-the-art review of the existing environmental regulations concerning every production sector involved in the construction process, identifying the implementation level of such regulations; (2) analysis of the existing environmental product declarations (EPDs) of construction products; (3) identify the different calculation methodologies of $CO_2$ emissions of construction processes; and finally, (4) the OERCO2 platform, which includes a software application for the calculation of $CO_2$ emissions of construction processes, for teaching purposes.

### Study of Environmental Regulations

The first objective was the study of environmental regulations in all sectors involved in construction and the level of implementation of these regulations in the countries participating in the project. This required collecting the regulations concerning the calculation of $CO_2$ emissions in the construction sector at European and national level of those countries that are involved in the project, generating Report 1 about regulations (see Fig. 1) in OERCO2 platform with hyperlinks to allow downloading the document. Each partner collected their own countries' regulations in their official language and in English, the last being the common language in the project.

In order to evaluate the degree of implementation of these regulations and the knowledge of the different agents regarding issues such as regulations, emission calculation tools, surveys were carried out in the different countries participating in the project. Figures 2 and 3 show an extract of that survey, and statistical studies on the results thereof. These results made it possible to determine the main targets to be tackled down in order to improve users' knowledge.

### EPDs in the Construction Sector

A study of EPDs in the construction sector of the participant countries was carried out. EPD report on environmental impacts such as, for example, the total kilograms of $CO_2$ equivalents generated in the manufacturing process of a product. This document can cover all phases of the life cycle, from the extraction of the raw material with which the product is manufactured until it is completely finished (UNE-EN ISO

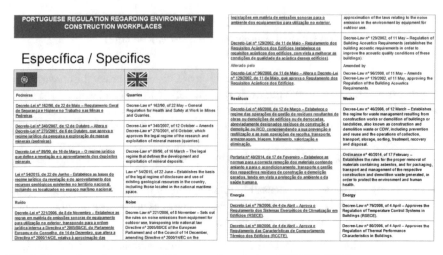

**Fig. 1** Extract from the report on environmental regulations (OERCO2 2018)

21930 2010). In order to perform this study, the project generated two reports: one on the most common materials used in building construction, analyzing the construction processes in each country and the corresponding materials employed in each process, and properly assessing their $CO_2$ emissions. The second report contains the environmental product declarations of each material included in the previous activity. Finally, a database was created which includes the construction materials and their carbon footprint, i.e., kilograms of $CO_2$ equivalents, generated from cradle to gate.

## Methodologies for the Calculation of $CO_2$ Emissions

The different methodologies for the calculation of the $CO_2$ emission in construction processes were studied in each participating country. Three reports were generated: (1) on the awareness of climate change and reduction of $CO_2$ emissions in universities (architecture and engineering schools), and organizations conducting EPDs; (2) a report on the methodologies for the calculation of $CO_2$ emissions per construction material; (3) a report about the possibilities for reuse or recycling of materials.

The first report was based on the results of a survey at universities (architectural and engineering) and organizations conducting EPDs, also asking professionals to verify the existing awareness of climate change and possible reduction of $CO_2$ emissions in this sector.

The manufacturing process of each material has certain phases and associated $CO_2$ emissions. The second report examined each of these phases and the emissions generated during its manufacture, to finally get the total number of kilograms of $CO_2$ equivalent. This analysis included emissions derived from the extraction of raw materials, their transportation, and manufacturing (cradle-to-gate), leaving transportation to the construction site out of scope.

The last report was about the possibilities for reuse or recycling of materials. All constituent materials of a building have a life cycle, and when it ends, such

**OERCO2**
CENTRO DE RECURSOS ONLINE PARA EL
ESTUDIO INNOVADOR DEL CICLO DE VIDA DE
LOS MATERIALES DE CONSTRUCCIÓN
2016-1-ES01-KA203-025422

Co-funded by the
Erasmus+ Programme
of the European Union

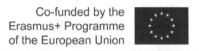

### SURVEY IN ACADEMIC WORLD

**Q1 What is your profession?**
- ❑ Professor
- ❑ Student
- ❑ Other _____

**Q2 What is your degree?**
- ❑ Architect
- ❑ Engineer
- ❑ Project Management
- ❑ Quantity Surveyor/Building Engineer
- ❑ Other _____

**Q3 In which country do you study/work?**
- ○ Spain
- ○ Italy
- ○ Portugal
- ○ Romania
- ○ Other _____

**Q4 How is the level or implementation on environmental aspects in your studies?**
- ○ None
- ○ Low
- ○ Medium
- ○ High

**Q5 About the following expertise areas, which of them it is possible to study in your university?**
- ○ Energy Efficiency
- ○ Environmental impact of materials
- ○ Waste management
- ○ Water management
- ○ Environmental regulations
- ○ Passive construction
- ○ Others_____

**Q6 According to your profession, how much influence do you think that you have over the selection of materials and construction products on a typical project?**
- ○ No influence
- ○ Little influence
- ○ Some influence
- ○ Strong influence
- ○ Primary influence

Consortium members: Universidad de Sevilla (US), Asociación Empresarial de Investigación Centro Tecnológico del Mármol, Piedra y Materiales (CTM), CertiMaC Soc. Cons. a r. L. (CertiMaC), Centro Tecnologico da Ceramica e do Vidro (CTCV), Universitatea Transilvania Din Brasov (UTBV), Asociatia Romania Green Building Council (RoGBC).

**Fig. 2** Extract from the survey made at academic level (OERCO2 2018)

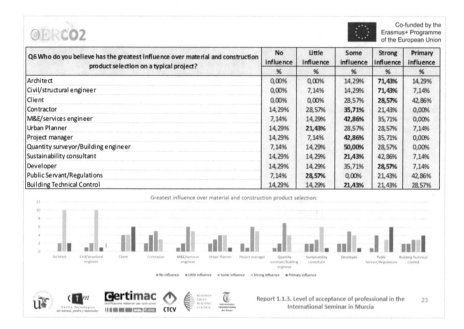

**Fig. 3** Extract from the statistical analysis of survey results (OERCO2 2018)

materials can be reused or recycled. In this report, a list was made of the possibilities of each material, with the emissions generated in each of these alternatives, in order to improve decision-making on the best option for the reuse of materials.

### Common European Curricula

One of the main aims of the project was to establish a common European curriculum for the specialization on the calculation of $CO_2$ emissions of construction works throughout the entire life cycle of buildings (construction–maintenance–deconstruction) in order to identify mitigation strategies. This curriculum is being implemented in two ways: cross-specialization in university careers AEC (Architecture, Engineering, and Construction) and continuous learning for professionals.

### Creation of the Platform OERCO2

The previous work is all integrated into the creation of the open-access OERCO2 platform and the software application for the calculation of $CO_2$ emissions in construction (Fig. 4). It is under constant revision by participants in the project, even after the project ends.

The curriculum developed in the previous section was implemented in a pilot specialization course through the OERCO2 Web site. The universities participating in the project (University of Seville and University of Transylvania, Brasov) implemented this pilot course, which evaluated and improved the curricula.

### Dissemination Activities

Dissemination of the project and the results obtained has been carried out through

**Fig. 4** OERCO2 platform, web main page (OERCO2 2018)

the contact network of the consortium. The impact of these dissemination activities has been continuously evaluated, monitoring the following indicators:

- Daily use of the online platform;
- Geographical origin of visits;
- The average time spent on the platform;
- Links from other Web pages to the OER of this project;
- Articles on press related to the project;
- The scientific articles dealing with the main idea of this project;
- The Google presence of the project;
- The presentation of the project in seminars, conferences, workshops, etc.

Among the main activities carried out were: international seminars (Murcia, Bucharest, and Seville), working meetings (Murcia, Faenza, Bucharest, and Seville), conferences and congresses, roundtables, brochures, and surveys.

- Representatives of the construction sector, higher education organizations, and authorities attending the seminars organized.
- A number of direct communications was sent to: local, regional, and national construction associations, higher education institutions and universities that teach courses or subjects related to construction, manufacture, or calculation of $CO_2$ emissions.

The results from this monitoring were useful to immediately make the appropriate decisions in terms of investment–result ratio.

Most of the indicators mentioned above have also been used to measure the project geographical impact.

These indicators are indirect because it is in the medium and long term when the project's main objective of reducing $CO_2$ emissions in the construction sector will be achieved.

In relation to deliver the activities, the project has the following strategies:

– The maintenance of the platform and the implementation of the project with free access to the training material. The main deliverable of the project was available as support material for classroom and online courses. This is the most important deliverable resource in terms of continuity and sustainability of the project. In addition, the update of news and events on the platform as well as relevant information regarding the OERCO2 teaching method for its future implementation in other European countries was promoted.
– The relevant reports and conclusions of the training development, the recommendations of higher education institutions and professionals of the sector, the opinions of the use of practical educational material, suggestions from experts, etc., were also included in the project platform after the end of the project in order to facilitate the dissemination of knowledge regarding $CO_2$ emissions and calculation methods in the community of the construction sector.
– The electronic bulletins of results were also distributed, which included relevant information on the actual impact observed on teachers and experts in the construction sector on topics related to the calculation of $CO_2$ emissions. The production and distribution of this electronic product were done once the project had finished, with the purpose of informing about the usefulness of the products beyond the European countries.
– The exploitation of the results is supported mainly by University of Seville, leader of the project, with international experience in the development of scientific articles on methodologies for calculating $CO_2$ emissions and with more than 76,000 enrolled students. University of Seville has published numerous articles on the subject addressed in this project and is also among the top four universities in Spain in terms of volume of results, so the importance of participation and dissemination is more than evident. It has counted with the collaboration of CTM as co-leader in the exploitation of results and also with the support of the other partners, especially for the dissemination strategies, the business networks which collaborates with the technology centers (Portugal, Italy) as well as the Green Building Council in Rumania.
– The wide dissemination carried out on the Internet also was an important way to support the future of the project, since once the application has a relevant number of visits, its maintenance and dissemination become much more interesting for the organizations involved in the project.

## 2.2 OERCO2 Software Tool

The OERCO2 tool (accessible at http://co2tool.oerco2.eu/) is an online application that enables the carbon footprint produced in the construction of residential buildings to be estimated. It stems from several previous research studies developed by the developers (Solís-Guzmán et al. 2013; González-Vallejo et al. 2015; Marrero et al. 2017) and includes the evaluation of $CO_2$ emissions for the construction process of

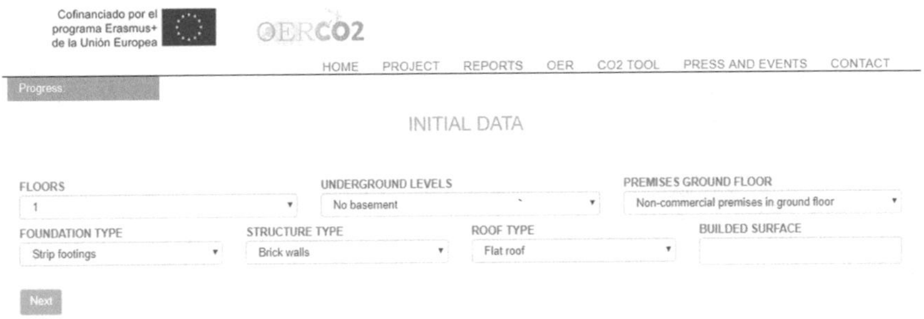

**Fig. 5** Selection interface for the initial data of the project

150 different residential building typologies. For the embodied energy assessment, the tool uses a cradle-to-site LCA analysis, that is, A1–A5 lifecycle phases, which correspond to manufacturing (A1–A3) and construction (A4–A5). The environmental data included in the OERCO2 tool was obtained from the Ecoinvent database through SimaPro. In order to obtain the $CO_2$ emissions embodied into construction materials, their lifecycle inventory (LCI) is analyzed by applying the IPCC 100a methodology, which is used by the carbon footprint indicator to isolate $CO_2$ and other GHG emissions from the LCI. The tool has been tested and evaluated by all the partners in the project, and it is considered that it includes all the building typologies and characteristics commonly used in Spain, Portugal, Italy, and Romania, which are essentially representative of those of any European country.

The methodology for the evaluation of the carbon footprint of the construction of residential buildings is based on the projects' bill of quantities and a classification system for construction works which breaks down this information on materials, manpower, and machinery necessary in the construction project. The budgets of 150 different projects are analyzed and classified; their budgets are reorganized in a construction-work breakdown system (CBS) that facilitates comparison. This organization system has been successfully applied in the previous research to estimate the generation of construction waste (Solís-Guzmán et al. 2009) and to evaluate the ecological footprint of buildings (Solís-Guzmán et al. 2013; González-Vallejo et al. 2015; Martínez-Rocamora et al. 2016b, 2017; Alba-Rodríguez et al. 2017).

The OERCO2 online tool is first divided into two stages: The first one refers to the basic characteristics of the project to be evaluated, called "initial data" (see Fig. 5), which requests information regarding the number floors above ground and basement that the building will have, the use that will be given to the ground floor (premises/dwelling), the type of foundation, structure, and roof. All these data will be selected through dropdowns for each item (see example in Fig. 6). In this same interface, the floor area is introduced, including basement square meters.

The basic characteristics of the building allow the software to look for the most similar project included, out of 150, in the tool's database. The matching project becomes the model for the calculation of quantities per floor square meter of all the items that will make up the project.

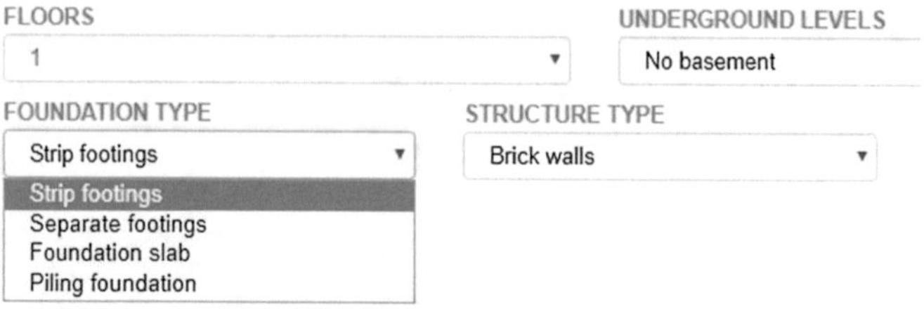

**FLOORS**

1

**UNDERGROUND LEVELS**

No basement

**FOUNDATION TYPE**

Strip footings

Strip footings
Separate footings
Foundation slab
Piling foundation

**STRUCTURE TYPE**

Brick walls

**Fig. 6** Foundation type in the tool

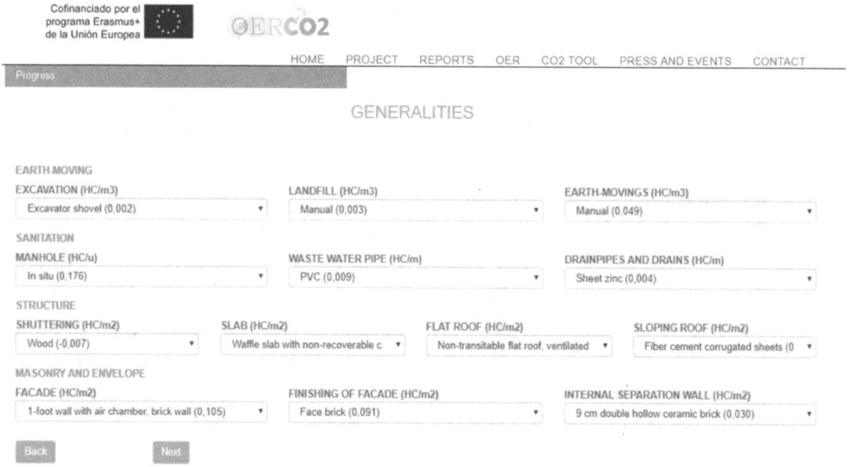

**Fig. 7** Selection interface for the generalities of the project

**Fig. 8** Selection interface for the characteristics of the project facilities

**Fig. 9** Selection interface for the finishing materials of the project

In the second stage, specific characteristics are selected, which are divided into three sections: generalities, (Fig. 7), installations (Fig. 8), and finishes (Fig. 9).

The first interface of this second stage is that of generalities (Fig. 7). In it, the characteristics for the items of earthworks, sewer system, structures, enclosures and building envelopes and interior partitions will be selected. For each of the mentioned items, the dropdown menu offers different constructive solutions, which are shown in Table 1 (Solís-Guzmán et al. 2018). Information on the environmental impact in carbon footprint (CF) of each available option, measured according to its specific measurement criteria, is provided in order to help in the decision-making of users, thus facilitating the selection of constructive solutions or materials with lower environmental impact for the construction of a project with minimal carbon emissions.

The second interface of this stage is that of installations (see Fig. 8). In it, the types of facilities and equipment included in the building are selected, as well as the materials. The available fields to be selected are air conditioning, heating, pipes' insulation, sanitary water supply, hot water production, and elevators. In the same way as in the previous interface, all the options included in the dropdown menus are shown in Table 1.

The third and last interfaces of this second stage are that of the finishes (see Fig. 9). Types of finishing materials are chosen for the following fields: insulation, cladding of walls, floors and roof, carpentry and protections. The available options are shown in Table 1.

To finish the analysis, the user must click on the "Compute" button. After a few seconds, the tool shows the last interface with a table containing the results obtained according to the selected options (see Fig. 10). These data are given in total and per $m^2$ of floor area. The data generated are:

**Table 1** Options available in each dropdown of the OERCO2 tool (Solís-Guzmán et al. 2018)

|  | Unit | Concept | Tool options available |
|---|---|---|---|
| INIT. INF. |  | *Number of floors* |  |
|  | N/A | Number of floors | 1/2/3/4/5/6+ |
|  | N/A | Number of underground floors | 0/1/2/3/4 |
|  | N/A | Shops in ground floor | No/Yes |
| C.02 |  | *Earthworks* |  |
|  | m³ | Excavations | Excavator/backhoe/not applicable |
|  | m³ | Fillings | Manual means/mechanic means/not applicable |
|  | m³ | Earth transport | Manual means/mechanic means/not applicable |
| C.03 |  | *Foundations* |  |
|  | m³ | Footings | Isolated/slab/strip/piles (m) |
| C. 04 |  | *Sewer system* |  |
|  | u | Manholes | In situ/prefabricated |
|  | m | Sewage pipes | PVC/concrete/fiber cement/polyethylene |
|  | m | Downpipes and roof sinks | Zinc sheet/steel sheet/reinforced PVC/polypropylene/fiber cement |
| C. 05 |  | *Structure* |  |
|  | m²/m³ | Supports | Brick wall/reinforced concrete |
|  | m² | Floor slabs | Waffle slab w/non-recoverable caissons/waffle slab w/recoverable caissons/one-way slab w/ceramic vaults/one-way slab w/concrete vaults/solid slab |
|  | m² | Formwork | Wood/metal |
| C. 06 |  | *Masonry* |  |
|  | m² | Façades | 1-ft brick wall w/chamber/1-/2-ft brick wall w/chamber/1-ft w/o chamber/1-/2-ft w/o chamber/1-ft w/chamber + plasterboard/1-/2-ft w/chamber + plasterboard/lightweight concrete block wall |

(continued)

**Table 1**  (continued)

|  | Unit | Concept | Tool options available |
|---|---|---|---|
|  | m² | Claddings | Ceramic brick/single-layer mortar/cement mortar/plastic paint/cement paint/ventilated cladding (natural stone/ceramic/resin/cellulose cement/wooden sandwich panel + XPS/cladding (artificial stone/limestone/marble/granite/wood) |
|  | m² | Partitions | Double hollow brick 9 cm/24 cm/triple hollow brick 15 cm/plasterboard |
| C. 07 |  | *Roof* |  |
|  | m² | Flat | Non-passable and ventilated/non-passable and non-ventilated/non-passable and inverted/passable and ventilated/passable and non-ventilated/passable and inverted/does not apply |
|  | m² | Sloping | Wavy fiber cement sheet/sandwich insulating panel/aluminum sheet/galvanized steel sheet/polyester/slate tiles/ceramic tiles/cement tiles/does not apply |
| C. 08 |  | *Installations* |  |
|  | u | Air conditioning system | Compact/parted system w/ducts/heat pump/VRF inverter/none |
|  | u | Terminal units | Ceiling unit/console/apartment type/none |
|  | m | Ducts | Glass fiber/galvanized steel/none |
|  | m | Pipes | Built-in galvanized steel/superficial galvanized steel/none |
|  | m² | Radiators | Classic steel/injected aluminum/iron/steel sheet/none |
|  | u | Boilers | Diesel/solid fuel/gas wall-mounted/mix electric wall-mounted/none |
|  | m | Cold water pipes | Copper/galvanized steel/polyethylene/polypropylene |
|  | m | Hot water pipes | Copper/galvanized steel/polypropylene |
|  | u | Sinks | PVC/polypropylene |
|  | m | Ventilation | Concrete/ceramic/helical galvanized Steel |
|  | u | Heater | Gas/electric/does not apply |
|  | u | Solar panels | Applies/does not apply |
|  | m | Pipe insulation | Applies/does not apply |

(continued)

**Table 1** (continued)

|  | Unit | Concept | Tool options available |
|---|---|---|---|
|  | u | Lift | Applies/does not apply |
| C. 09 |  | *Insulation* | |
|  | m² | Thermal acoustic | Polystyrene/polyurethane/glass fiber/rock wool/perlite/cork/polyethylene/none |
| C. 10 |  | *Finishes* | |
|  | m² | Continuous claddings | Gypsum plaster/cement mortar/does not apply |
|  | m² | Floorings | Ceramic/stoneware/continuous concrete/hydraulic tile/linoleum/carpet/cork/softwood parquet/floating solid softwood/floating laminated softwood/hardwood parquet/floating solid hardwood/floating laminated hardwood/limestone/marble/slate/granite/terrazzo/concrete slab |
|  | m² | Ceilings | Continuous plaster w/rods/continuous plaster w/metal fixings/removable plaster panels/continuous laminated gypsum/removable laminated gypsum w/hidden support grid |
| C. 11 |  | *Carpentry and protection elements* | |
|  | m² | Windows | Pine wood casement/lacquered aluminum sliding/lacquered aluminum casement w/thermal bridge break/PVC sliding |
|  | m² | Doors | Wood/melamine |
|  | m² | Blinds | Anodized aluminum/PVC/wood/none |
|  | m² | Protection grids | Hot-rolled steel/none |
|  | m | Railings | Steel/aluminum/wood/none |
| C. 12 |  | *Glass and synthetics* | |
|  | m² | Glazing | Thermal acoustic $6 + 12 + 6/6 + 12 + 6$ low-emissive/$8 + 14 + 5$/$5+5$ low-emissive argon and solar control |
| C. 13 |  | *Paintings* | |
|  | m² | Exterior | Plastic paint/cement paint/does not apply |
|  | m² | Interior | Plastic paint/does not apply |

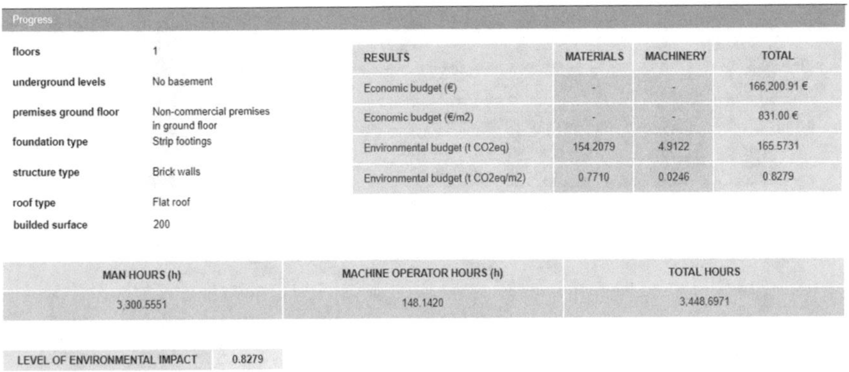

| Progress | | | | | |
|---|---|---|---|---|---|
| floors | 1 | RESULTS | MATERIALS | MACHINERY | TOTAL |
| underground levels | No basement | Economic budget (€) | - | - | 166,200.91 € |
| premises ground floor | Non-commercial premises in ground floor | Economic budget (€/m2) | - | - | 831.00 € |
| foundation type | Strip footings | Environmental budget (t CO2eq) | 154.2079 | 4.9122 | 165.5731 |
| structure type | Brick walls | Environmental budget (t CO2eq/m2) | 0.7710 | 0.0246 | 0.8279 |
| roof type | Flat roof | | | | |
| builded surface | 200 | | | | |

| MAN HOURS (h) | MACHINE OPERATOR HOURS (h) | TOTAL HOURS |
|---|---|---|
| 3,300.5551 | 148.1420 | 3.448.6971 |

| LEVEL OF ENVIRONMENTAL IMPACT | 0.8279 |
|---|---|

**Fig. 10** Interface of the results of the evaluation of the project

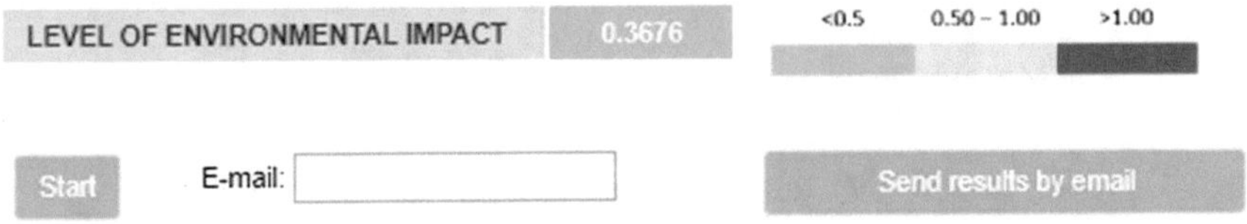

| LEVEL OF ENVIRONMENTAL IMPACT | 0.3676 | <0.5 | 0.50 – 1.00 | >1.00 |
|---|---|---|---|---|

| Start | E-mail: | Send results by email |
|---|---|---|

**Fig. 11** Scale of level of environmental impact for the projects evaluated

- Economic budget of the construction project analyzed (in € and €/m$^2$). This result is the sum of the materials and machinery costs.
- Environmental budget of the construction project analyzed (t $CO_2$-eq and t $CO_2$-eq/m$^2$). This result is the sum of the $CO_2$ emissions of materials and machinery.
- Total labor hours necessary to carry out the construction.
- Total machinery hours necessary for the construction.

The data to generate the results of the software tool come from:

- The amounts of each item per floor area, established by the selected project type based on the initial data set by the user;
- The built surface of the project that is being evaluated;
- The internal construction cost database of the software tool, based on the Andalusia Construction Costs Database (ACCD) (Andalusia Government 2017), where resources are collected for each work unit, as well as its cost. The CF information has been added to ACCD by the authors, obtained from the databases of Ecoinvent and SimaPro.

The level of environmental impact of the project analyzed is provided, so that the user can easily visualize the environmental impact of the project (see Fig. 11). These environmental impact levels have been established according to the analysis compiled in Chastas et al. (2018). The user can also receive by email the results obtained from the software tool.

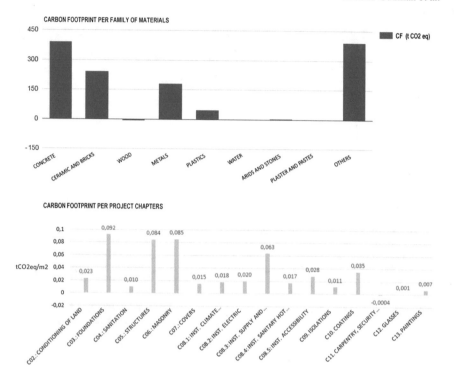

**Fig. 12** CF (t $CO_2$-eq) per family of materials and per project chapters

Finally, the tool provides two graphs; see Fig. 12, which details the environmental impact on CF per family of materials and per chapters of the project. This information helps to the user to visualize the main sources of the impact of his project.

## 3 Case Studies

The CF of different construction typologies was analyzed in order to determine the tool sensitivity to changes in the project characteristics. The projects studied have 2, 4, 5 and more than 5-story buildings with different types of foundations (reinforced slab, isolated footing or pile foundation). All of them lack of underground floors and premises on the ground floor, and all have a reinforced concrete structure and horizontal roofs (see Table 2 for other project characteristics).

**Table 2** Case studies combinations

| Code | No. of storeys | Type of foundation |
| --- | --- | --- |
| c008 | Two | Isolated footing |
| c112 | Two | Reinforced slab |
| c116 | Two | Pile foundation |
| c004 | Four | Isolated footing |
| c060 | Four | Pile foundation |
| c012 | Five | Isolated footing |
| c062 | Five | Pile foundation |
| c014 | + Five | Isolated footing |
| c063 | + Five | Pile foundation |

# 4   Results

Firstly, the foundation impact is analyzed for each typology of building, and secondly, different construction materials are proposed to study its impact.

In 2-story buildings, the project with reinforced concrete foundations has the lowest total CF 0.57 (all the results of CF are expressed in t $CO_2$ eq) as compared to that with pile foundations, which has 0.67 CF total/m$^2$ (Fig. 13). This is due to the influence of the other project characteristics, because in construction projects the foundation is selected based on the soil quality, which affects not only the calculation and design, but also the structure and earthworks. The CF of the type of foundation when analyzed in isolation is the lowest for both, the reinforced concrete slab and the isolated footing (Fig. 13).

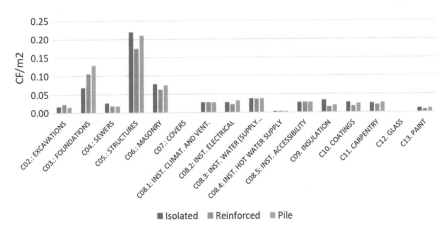

**Fig. 13** CF generated by chapters in the project, 2-story with different types of foundation

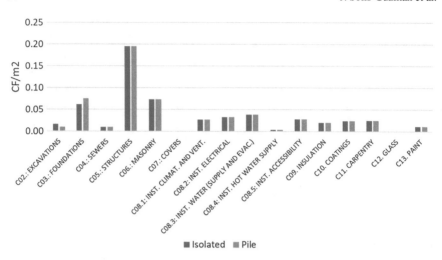

**Fig. 14** CF per budget chapter of 4-story building with different types of foundation

Improvements in the project by making changes in the structure, thermal insulation, and cladding can reduce 15% of the total CF. Table 3 summarizes the alternative solutions and the original impact.

In four- and five-floor buildings, the total CF/m$^2$ obtained in the different projects does not vary, meaning that the type of foundation chosen (isolated footing or pile foundation) does not influence the total CF/m$^2$ (Figs. 14 and 15). This is because the foundation and excavation chapters' impacts are compensated. The isolated footing has a smaller footprint than the pile foundation due to the lower amount of reinforcement needed. But the previous earthwork needed in the pile foundation has a lower impact due to the employment of more efficient machines and that less soil is excavated than in isolated footing buildings. In Figs. 14 and 15, each chapter of the project budget is represented. The only significant difference observed between projects is in the foundations' chapters.

The total CF is reduced 18% percent in the project with isolated footing foundations when the improvements proposed in Table 3 are implemented in the structure, thermal insulation, and coatings. Figure 16 represents the reduction of CF per chapter.

**Table 3** Proposed improvements to the projects

| Element | Initial characteristics | CF/m$^2$ | Improvements | CF/m$^2$ |
|---|---|---|---|---|
| Structure | Waffle slab with permanent coffers | 0.110 | Slab with concrete blocks | 0.056 |
| | Metallic formwork | 0.004 | Wood formwork | −0.007 |
| Insulation | Polyethylene | 0.025 | Mineral wool | 0.012 |
| Floor coating | Marble | 0.005 | Solid wood flooring | −0.012 |

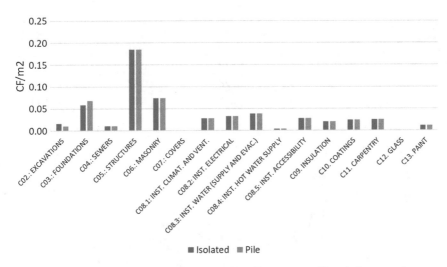

**Fig. 15** CF per budget chapter 5-story projects with different types of foundation

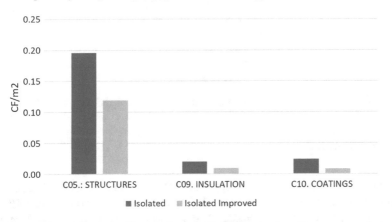

**Fig. 16** Reduction of CF obtained in each chapter with the introduction of improvements in 4-story buildings

However, the impact on the cost per m², as a consequence of the improvements proposed, only increases of 1%.

In +5-story buildings, the lowest CF is obtained with the pile foundation. This result differs to that obtained for 4- and 5-story buildings. This is due to the impact of the pile length on the total CF, which is not proportional to the height of the building but, otherwise, it has a lower impact per square meter of floor constructed. The Qi of "in situ piles" is reduced to more than half with respect to 4- and 5-story buildings (see Table 4) (Fig. 17).

**Table 4** Qi of the reinforcement steel (kg) per floor square meter in each project

| Storeys | Steel (t/m of pile) | In situ pile (m/floor m$^2$) | Proportion |
|---|---|---|---|
| 4 | 5.19 | 0.36 | 14.41 |
| 5 | 4.78 | 0.32 | 14.93 |
| +5 | 4.36 | 0.14 | 31.14 |

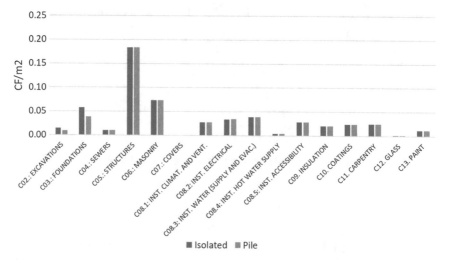

**Fig. 17** CF per budget chapter in more than 5-story buildings with different types of foundation

## 5 Conclusions

The present work proposes a teaching tool for the introduction of $CO_2$ calculations in construction projects for university students as part of the work in the OERCO2 Erasmus+ project. The environmental impact assessment of buildings from early stages of design becomes essential for improving their sustainability. For this, carbon footprint calculations allow to determine the emissions of greenhouse gases derived from the building process from cradle to site.

The main aims of the OERCO2 project have been presented, the last of which consists of the development of an online tool for the CF assessment of the construction of residential buildings.

This tool allows users to detect opportunities, both environmental and economic, of the improvement of their projects through modifications such as the selection of different materials or constructive solutions.

Regarding results obtained, the CF of different construction typologies has been analyzed in order to determine the tool sensitivity. Projects with two floors and pile foundation have the highest impact of all cases studied, and those with more than five floors and a pile foundation have the lowest CF per floor area.

The study also shows that the chapters that control the total CF are foundations, structures, and masonry. The structures and foundations chapters have a higher CF per floor square meter in 2-story buildings.

In the case of 2-story buildings, the pile foundation generates the highest CF. However, different foundations analyzed have the same impact on the total CF/m².

CF reductions can be achieved by making changes in the materials used, which meets the main goal of the OERCO2 project in introducing $CO_2$ calculations in engineering and architecture curricula in order to reduce from the design phase the environmental impact of future projects.

**Acknowledgements** This research has been funded by the OERCO2 project (code 2016-1-ES01-KA203-025422), an ERASMUS + project co-funded by the European Union and within the framework of an initiative of year 2016 (KA2, Strategic Partnerships in the field of higher education), with the support of the Servicio Español Para la Internacionalización de la Educación (SEPIE, Spain).

# References

Alba-Rodríguez MD, Martínez-Rocamora A, González-Vallejo P et al (2017) Building rehabilitation versus demolition and new construction: economic and environmental assessment. Environ Impact Assess Rev 66:115–126. https://doi.org/10.1016/J.EIAR.2017.06.002

Andalusia Government (2017) Andalusia Construction Cost Database (ACCD). https://www.juntadeandalucia.es/organismos/fomentoyvivienda/areas/vivienda-rehabilitacion/planes-instrumentos/paginas/bcca-sept-2017.html. Accessed 11 Jan 2019

Asociación Ecómetro (2017) Ecometro LCA tool website. http://acv.ecometro.org/. Accessed 30 Dec 2018

Bare JC, Hofstetter P, Pennington DW, Haes HaU (2000) Midpoints versus endpoints: the sacrifices and benefits. Int J Life Cycle Assess 5:319–326. https://doi.org/10.1007/bf02978665

BREEAM (2017) BREEAM ES website. http://www.breeam.es/. Accessed 30 Dec 2018

Buyle M, Braet J, Audenaert A (2013) Life cycle assessment in the construction sector: a review. Renew Sustain Energy Rev 26:379–388. https://doi.org/10.1016/j.rser.2013.05.001

Cabeza LF, Rincón L, Vilariño V et al (2014) Life cycle assessment (LCA) and life cycle energy analysis (LCEA) of buildings and the building sector: a review. Renew Sustain Energy Rev 29:394–416. https://doi.org/10.1016/j.rser.2013.08.037

Cagiao J, Gómez B, Doménech JL et al (2011) Calculation of the corporate carbon footprint of the cement industry by the application of MC3 methodology. Ecol Indic 11:1526–1540. https://doi.org/10.1016/j.ecolind.2011.02.013

Chastas P, Theodosiou T, Kontoleon KJ, Bikas D (2018) Normalising and assessing carbon emissions in the building sector: a review on the embodied $CO_2$ emissions of residential buildings. Build Environ 130:212–226

Chau CK, Leung TM, Ng WY (2015) A review on life cycle assessment, life cycle energy assessment and life cycle carbon emissions assessment on buildings. Appl Energy 143:395–413. https://doi.org/10.1016/j.apenergy.2015.01.023

Chen Q (2010) Use of open educational resources: challenges and strategies. Lecture notes in computer science (including subseries Lecture notes in artificial intelligence and lecture notes in bioinformatics), pp 339–351

Clements KI, Pawlowski JM (2012) User-oriented quality for OER: understanding teachers' views on re-use, quality, and trust. J Comput Assist Learn 28:4–14. https://doi.org/10.1111/j.1365-2729.2011.00450.x

Cype Ingenieros SA (2018) Cypetherm HE PLus. https://bimserver.center/bim_store.asp. Accessed 30 Dec 2018

Di Blas N, Fiore A, Mainetti L et al (2014) A portal of educational resources: providing evidence for matching pedagogy with technology. Res Learn Technol 22. https://doi.org/10.3402/rlt.v22.22906

Dossche C, Boel V, De Corte W (2017) Use of life cycle assessments in the construction sector: critical review. Procedia Eng 171:302–311. https://doi.org/10.1016/j.proeng.2017.01.338

e2CO2cero (2014) e2CO2cero tool website. http://tienda.e2co2cero.com/. Accessed 30 Dec 2018

Ecoinvent Centre (2013) Ecoinvent database v3 ecoinvent report. www.ecoinvent.org. Accessed 30 Dec 2018

European Parliament—Council of the European Union (2018) Directive (EU) 2018/844 of the European Parliament and of the Council of 30 May 2018 amending Directive 2010/31/EU on the energy performance of buildings and Directive 2012/27/EU on energy efficiency (Text with EEA relevance). Off J Eur Union 61:75. https://doi.org/10.1007/3-540-47891-4_10

Geng S, Wang Y, Zuo J et al (2017) Building life cycle assessment research: a review by bibliometric analysis. Renew Sustain Energy Rev 76:176–184. https://doi.org/10.1016/j.rser.2017.03.068

Giesekam J, Barrett JR, Taylor P (2016) Construction sector views on low carbon building materials. Build Res Inf 44:423–444. https://doi.org/10.1080/09613218.2016.1086872

González-Vallejo P, Marrero M, Solís-Guzmán J (2015) The ecological footprint of dwelling construction in Spain. Ecol Indic 52:75–84. https://doi.org/10.1016/j.ecolind.2014.11.016

ISO 15686-5 (2017) Buildings and constructed assets—service life planning—Part 5: life-cycle costing

Khanna P, Basac PC (2013) An OER architecture framework: needs and design. Int Rev Res Open Distrib Learn 14:65–83. https://doi.org/10.11648/j.ijnfs.20140304.26

Llamas-Nistal M, Mikic-Fonte FA (2014) Generating OER by recording lectures: a case study. IEEE Trans Educ 57:220–228. https://doi.org/10.1109/TE.2014.2336630

Marrero M, Martin C, Muntean R et al (2018) Tools to quantify environmental impact and their application to teaching: projects City-zen and HEREVEA. IOP Conf Ser Mater Sci Eng 399:012038. https://doi.org/10.1088/1757-899X/399/1/012038

Marrero M, Puerto M, Rivero-Camacho C et al (2017) Assessing the economic impact and ecological footprint of construction and demolition waste during the urbanization of rural land. Resour Conserv Recycl 117. https://doi.org/10.1016/j.resconrec.2016.10.020

Martínez-Rocamora A, Solís-Guzmán J, Marrero M (2016a) LCA databases focused on construction materials: a review. Renew Sustain Energy Rev 58:565–573. https://doi.org/10.1016/j.rser.2015.12.243

Martínez-Rocamora A, Solís-Guzmán J, Marrero M (2016b) Toward the ecological footprint of the use and maintenance phase of buildings: utility consumption and cleaning tasks. Ecol Indic 69:66–77. https://doi.org/10.1016/j.ecolind.2016.04.007

Martínez Rocamora A, Solís-Guzmán J, Marrero M (2017) Ecological footprint of the use and maintenance phase of buildings: maintenance tasks and final results. Energy Build 155. https://doi.org/10.1016/j.enbuild.2017.09.038

McGreal R, Sampson DG, Chen NS et al (2012) The open educational resources (OER) movement: free learning for all students. In: ICALT 2012—12th IEEE international conference on advanced learning technologies, pp 748–751

Nipa TJ, Kermanshachi S (2018) Analysis and assessment of graduate students' perception and academic performance using open educational resource (OER) course materials. In: ASEE 2018—annual conference and exposition

OERCO2 (2018) OERCO2—Construction material life cycle website. In: Erasmus + la Unión Eur. http://oerco2.eu/. Accessed 1 Jan 2019

PRé Sustainability (2016) SimaPro 8. https://simapro.com/. Accessed 28 Mar 2018

Ramesh T, Prakash R, Shukla KK (2010) Life cycle energy analysis of buildings: an overview. Energy Build 42:1592–1600. https://doi.org/10.1016/j.enbuild.2010.05.007

Schwartz Y, Raslan R, Mumovic D (2018) The life cycle carbon footprint of refurbished and new buildings—a systematic review of case studies. Renew Sustain Energy Rev 81:231–241. https://doi.org/10.1016/j.rser.2017.07.061

SEPIE (2017) Información Educacion Superior 2017—Servicio Español para la Internacional-ización de la Educación. http://sepie.es/educacion-superior/informacion-2017.html. Accessed 4 Jan 2019

SOFIAS Project (2017) SOFIAS project website. http://www.sofiasproject.org/. Accessed 30 Nov 2018

Solís-Guzmán J, Marrero M, Montes-Delgado MV, Ramírez-de-Arellano A (2009) A Spanish model for quantification and management of construction waste. Waste Manag 29:2542–2548. https://doi.org/10.1016/j.wasman.2009.05.009

Solís-Guzmán J, Marrero M, Ramírez-De-Arellano A (2013) Methodology for determining the ecological footprint of the construction of residential buildings in Andalusia (Spain). Ecol Indic 25. https://doi.org/10.1016/j.ecolind.2012.10.008

Solís-Guzmán J, Rivero-Camacho C, Alba-Rodríguez D, Martínez-Rocamora A (2018) Carbon footprint estimation tool for residential buildings for non-specialized users: OERCO2 project. Sustain 10. https://doi.org/10.3390/su10051359

Spain METDA (2017) Simplified procedures for building energy certification. http://www.minetur.gob.es/ENERGIA/DESARROLLO/EFICIENCIAENERGETICA/CERTIFICACIONENERGETICA/DOCUMENTOSRECONOCIDOS/Paginas/procedimientos-certificacion-proyecto-terminados.aspx. Accessed 30 Nov 2018

Spain Ministry of Development (2015) Unified Tool LIDER-CALENER (HU-Tool). http://www.codigotecnico.org/web/recursos/aplicaciones/contenido/texto_0004.html. Accessed 30 Dec 2018

SpainGBC (2015) LEED certificate. http://www.spaingbc.org/web/leed-4.php. Accessed 30 Nov 2018

SpainGBC (2013) VERDE tool website. http://www.gbce.es/es/pagina/herramientas-de-evaluacion-de-edificios. Accessed 30 Nov 2018

Tovar E, Lopez-Vargas JA, Piedra NO, Chicaiza JA (2013) Impact of open educational resources in higher education institutions in Spain and Latin Americas through social network analysis. In: 120th ASEE annual conference and exposition

Tovar E, Piedra N (2014) Guest editorial: open educational resources in engineering education: various perspectives opening the education of engineers. IEEE Trans Educ 57:213–219. https://doi.org/10.1109/TE.2014.2359257

Tuomi I (2013) Open educational resources and the transformation of education. Eur J Educ 48:58–78. https://doi.org/10.1021/acs.orglett.8b00518

UNE-EN 15804 (2012) Sustainability of construction works—environmental product declara-tions—core rules for the product category of construction products

UNE-EN 15978 (2012) Sustainability of construction works. Assessment of environmental perfor-mance of buildings. Calculation method

UNE-EN ISO 14001 (2015) Environmental management systems—requirements with guidance for use

UNE-EN ISO 14020 (2002) Environmental labels and declarations—general principles

UNE-EN ISO 14021 (2017) Environmental labels and declarations—self-declared environmental claims (Type II environmental labelling)

UNE-EN ISO 14025 (2006) Environmental labels and declarations—type III environmental decla-rations—principles and procedures

UNE-EN ISO 14040 (2006) Environmental management—life cycle assessment—principles and framework

UNE-EN ISO 14044 (2006) Environmental management—life cycle assessment—requirements and guidelines

UNE-EN ISO 21930 (2010) Sustainability in building construction—environmental declaration of building products

Weidema BP, Thrane M, Christensen P et al (2008) Carbon footprint: a catalyst for life cycle assessment? J Ind Ecol 12:3–6. https://doi.org/10.1111/j.1530-9290.2008.00005.x

Zancanaro A, Todesco JL, Ramos F (2015) A bibliometric mapping of open educational resources. Int Rev Res Open Distrib Learn 16:1–23. https://doi.org/10.1063/1.527861

# Carbon Footprints of Agriculture Sector

Bhavna Jaiswal and Madhoolika Agrawal

**Abstract** Climate change being today's major issue is concerned with the unprecedented increase in natural resource exploitation and uncontrolled population increase, reaching in an irreversible point. Greenhouse gases (GHGs) responsible for such changes are emitted by a variety of natural as well as anthropogenic sources. Agriculture sector shares a major proportion in total GHG emission. As the food demand is increasing with the rising population, the proportion of GHG emissions from agricultural sector is also increasing. The total amount of GHGs (in terms of carbon equivalent (C-eq)) emitted by the processes in agricultural sector is regarded as carbon footprint of agriculture. Various activities related to agriculture such as plowing, tilling, manuring, irrigation, variety of crops, rearing livestock, and related equipment emit a significant amount of GHGs that are categorized in three tiers of carbon footprinting, separated by hypothetical boundaries. The energy input through machinery, electricity, livestock management, and fossil fuel constitutes a major proportion of carbon emission through agriculture. Crop cultivation system mainly cereals produces higher GHGs than any other farming systems like vegetables and fruits. Beside this, land-use changes including conversion of natural ecosystem to agricultural, deforestation, and crop residue burning after harvest contribute significantly to higher carbon emission. This review article will focus on carbon footprint from agriculture including inputs for uses from energy, fertilizers, organic manure, pesticides, and processes that affect carbon emission from agriculture. The mitigation practices effective in reducing the carbon footprinting from various agricultural activities will also be reviewed. Efficient use of fossil fuel and other non-renewable energy sources in the agriculture system, diversified cropping system, enhancing soil carbon sequestration by straw return, plantation, etc., crop rotation system, and limiting deforestation will be discussed as measures which may help to reduce the GHG emissions from agriculture sector.

**Keywords** Greenhouse gas · Agriculture · Carbon footprint · Mitigation strategies · Farming practices

B. Jaiswal · M. Agrawal (✉)
Laboratory of Air Pollution and Global Change, Department of Botany,
Institute of Science, Banaras Hindu University, Varanasi 221005, India
e-mail: madhoo.agrawal@gmail.com

© Springer Nature Singapore Pte Ltd. 2020
S. S. Muthu (ed.), *Carbon Footprints*, Environmental Footprints and Eco-design
of Products and Processes, https://doi.org/10.1007/978-981-13-7916-1_4

# 1  Introduction

Climate change is a major concern for society, as it causing shift in weather patterns such as unpredictable precipitation, extreme temperatures, higher occurrence of flood, drought and cyclones. To prevent the extreme weather variables and spread awareness, various inventories are prepared. Among them, the term carbon footprint has become a widely discussed term as the planet has witnessed the effects of climate change. The concept of carbon footprint is taken from ecological footprint given by Rees in 1992. Ecological footprint can be defined as biologically productive land and sea area required to sustain a given human population, expressed in terms of global hectares. Likewise, Wiedmann and Minx (2008) defined carbon footprint as a certain amount of gaseous emission that is relevant to climate change and associated with human production or consumption activity. Carbon footprint is thus emission of greenhouse gases (GHGs) from all sources and processes related to a particular product or individual or system, from manufacturing to disposal. Earlier, only $CO_2$ is taken under consideration for carbon footprint estimation, but at present, all the major GHGs emitted such as $CO_2$, $CH_4$, and $N_2O$ are taken under consideration in terms of $CO_2$ equivalent ($CO_2$-e). IPCC (2014) has given the definition of $CO_2$-e as $CO_2$ concentration that would cause the same radiative forcing as a given mixture of $CO_2$ and other forcing components.

Carbon footprint is a component of life-cycle assessment (LCA) that measures GHGs, whereas LCA assesses all the environmental impact associated with a product. Carbon footprint is calculated by dividing the whole process tier-wise, separated by hypothetical boundaries. Global warming potential (GWP) of all tiers adds to carbon footprint. Since in agriculture system, no standards are available, so boundaries are not decided. Tier 1 consists of all the direct on-site emission, such as emission from soil and machinery. Tier 2 consists of indirect farm emissions such as from electricity, and tier 3 involves all the indirect emissions related to manufacturing and transport of agriculture-based chemicals and machinery, etc. Pandey and Agrawal (2014) have given the formula to calculate carbon footprint.

GWP of tier (kg $CO_2$-e ha$^{-1}$) = emission/removal of $CH_4$ × 25 + emission/removal of $N_2O$ × 298 + emission/removal of $CO_2$

Carbon footprint = $\sum$ (GWP of all tiers)

Carbon footprint from agriculture is calculated by the following formula (Lal 2004).

Carbon footprint

$$= \left( \sum \text{Agricultural input} * \text{GHG emission coefficients} \right) / (\text{Grain yield})$$

Agriculture is one of the major contributors in total GHG emissions. The processes related to agricultural practices from industry to farm to house, emit GHGs at every step that significantly contributes in global warming. With the increasing population demand, use of chemicals, electrical energy, and use of fossil fuels are the primary sources of emission from agriculture. The rate of natural resources exploitation in terms of fossil fuel use, minerals, and carbon utilization from soil, etc., is much higher, than the input to the soil as carbon sequestration. Chemical fertilizers and pesticides are available at a cheap rate, and to increase the productivity, these are utilized in the farming system, on the cost of deterioration of natural resources. With due course of time, soil is losing its property to foster life due to various anthropogenic activities such as deforestation, erosion, use of chemicals, and disposal of hazardous wastes. This review focuses on aspects of carbon footprint from agriculture sector, including emission from pre-farm, on-farm, and post-farm activities. Various mitigation strategies regarding farming practices are suggested, and the models for footprint estimation are discussed.

## 2   Methodology

For literature survey, the World Wide Web was searched for relevant information. Google Scholar, PubMed, and ResearchGate were used for finding papers using keywords such as carbon footprint from agriculture, mitigation of greenhouse gas emission, emission from agriculture, fertilizer, machinery, electricity, livestock contribution in carbon footprint, estimation of carbon footprint, models for estimation of carbon footprint, etc., and 150 papers were finally selected. The articles published from 2010 to February 2019 were considered. Data of Food and Agricultural Organization (FAO), Intergovernmental Panel on Climate Change (IPCC), and International Federation of Organic Agriculture Movements (IFOAM) were also used.

## 3   Components of Agriculture and Their Contribution in Carbon Footprint

Emission from total agricultural sector has increased since records (FAOSTAT 2019). Agriculture sector along with land-use change accounts for one-fourth of total anthropogenic GHG emission (IPCC 2014). World population has increased by 36%, and the agricultural land has increased by 42.5% since 1990 to 2014, but emission from agriculture, forestry, and land use has increased by 1.1% (FAO 2015). Emission from India has increased by 11.8% from agriculture, forestry, and land use, whereas the increase in population and harvested area was increased by 45.8 and 50.8%, respectively (FAO 2015). Asia has a major share in total emission from agriculture that is 44%, followed by America, Africa, Europe, and Oceania (FAO 2014). Various factors responsible for carbon footprint in different tiers are shown in Fig. 1.

**Total energy**: Yuosefi et al. (2017) calculated the carbon footprint of sunflower cultivation in Iran. It was reported that sunflower cultivation requires 70.31% direct energy that involves human labor, diesel, water for irrigation and electricity, and 29.69% indirect energy that involves seeds, fertilizers, pesticides, and machinery. In terms of renewable energy, 20.97% include human labor, seeds, and water for

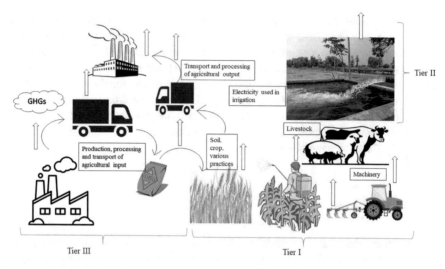

**Fig. 1** Emission sources from agriculture sector under different tiers

irrigation and 79.03% of non-renewable energy includes diesel, pesticides, fertilizers, electricity, and machinery (Yousefi et al. 2017). Energy from fossil fuel that is utilized in agriculture contributes maximally in GHG emissions (Yadav et al. 2018). From the utilization of energy in agriculture using fossil fuel, 785 million tons of $CO_2$-e was emitted globally in the year 2010 (FAO 2014). Crop residue removal after harvest utilizes energy and thus increases the carbon footprint by 6% per hectare (Goglio et al. 2014).

Fossil fuel utilization in agriculture emits GHGs, and the production and transport of fossil fuel also add to the emission of GHGs. From livestock sector, energy uses contributed 20% of the total emission from livestock sector (Gerber et al. 2013). Input of energy contributed only 8% in semiarid areas compared to other agricultural inputs such as fertilizers and pesticides which contributed 82 and 9%, respectively (Devakumar et al. 2018). Pre-farm processes such as production of fertilizers, pesticides, diesel, and electricity in cotton production in Australia contributed 25.3%, whereas post-farm processes such as cottonseed drying utilizing LPG, gin machinery utilizing electricity, bale packaging, gin trash treatment, and transportation accounted for 26.2% (Hedayati et al. 2019). Rest 48.4% were emitted during on-farm processes.

Electricity used in agriculture contributed the most in carbon footprinting (Yousefi et al. 2017). Yousefi et al. (2017) reported that the contribution of electricity is 78.7% of carbon footprinting. Electricity used for irrigation in rice farming in China contributed 4% of total carbon footprinting, whereas for wheat and maize farming contributed 37 and 18%, respectively (Zhang et al. 2017). Sah and Devakumar (2018) reported 3% emission from the use of electricity from India for the period of 2000–2010. In soybean oil production in Thailand, electricity share was 6% of total carbon footprint (Patthanaissaranukool and Polprasert 2016).

Irrigation contribution is small in GHG emission that is 1–13%, except for wheat and rice in India (Rao et al. 2019). Carbon footprint of rainfed agriculture is lower than the irrigated areas as the emission related to irrigation is reduced and the areas are smaller, so the practices are done manually (Devakumar et al. 2018). Irrigation of cotton cropping contributed 13.9% of the total on-farm carbon footprint (108 kg $CO_2$e $t^{-1}$ lint) (Hedayati et al. 2019).

Lal (2004) reported land-use change from natural to agricultural land causes the reduction in SOC by 60–75%. Low SOC not only reduces the productivity of plants, but also reduces the nutrient use efficiency of the plants as well as sequestration of atmospheric carbon (Lal 2011). SOC also decides the N status of the soil (Nath et al. 2017). Deforestation causes the emission of 0.81 Pg C $year^{-1}$ during 2000–2005 that contributed 7–14% of total emission during that time period (Harris et al. 2012). Conversion of forest land to agricultural land or pastures accounted 6–17% in total global GHG emission (IFOAM 2016).

**Machinery**: For the growing demand of population, the use of machinery is increasing and the energy provided to machinery is fulfilled by fossil fuel that is responsible for GHG emission. For production, transportation, and application, the fossil fuels are used extensively. In sunflower production, machinery contributed 67.176 kg $CO_2$-e $ha^{-1}$ out of 2042.091 kg $CO_2$-e $ha^{-1}$ total $CO_2$-e (3.29%) emissions (Yousefi et al. 2017). Rice contribution in total carbon footprint was reported to be 13% from machinery using fossil fuels in China (Zhang et al. 2017). Zhang et al. (2017) reported 25 and 20% emission from fuel utilization by machinery in wheat and maize cropping, respectively. For production of cotton in Australia, Hedayati et al. (2019) calculated carbon footprint of machinery and found it to account for almost 7% (125.5 kg $CO_2$e $t^{-1}$ lint) in total carbon footprint including pre-farm, on-farm, and post-farm processes. Among the three processes, the contribution of on-farm emission was 16% from machinery. Farag et al. (2013) estimated the carbon footprint of rice production in Egypt and found the lowest contribution of machinery in total emission.

**Diesel**: Diesel consumption occurs in transport of fertilizers, pesticides, seeds, and other farm equipment, and major emissions occur during tillage process. The consumption of diesel is dependent upon the size of tractor, tillage depth, frequency, and type of tillage. Yousefi et al. (2017) reported that the diesel consumption contributes to 12.24% carbon footprint during sunflower production. In conventional tillage and no-tillage system in rice–mustard cropping, diesel contributed 19 and 6%, respectively, in total carbon footprint (Yadav et al. 2018). From India, for the period of 2000–2010, contribution of diesel in carbon footprint was estimated less than 1% by Sah and Devakumar (2018). Patthanaissaranukool and Polprasert (2016) reported 38% contribution of diesel in soybean oil production from Thailand that accounts for 270 kg $CO_2$-e $ton^{-1}$ soybean oil and 17% contribution was recorded from heavy oil.

**Chemicals**: Fertilizer production, transport, and application contribute significantly in total GHG emission (Rao et al. 2019). Synthetic fertilizers account for 13% in total agricultural emission of GHGs (FAO 2014). In production process, major GHG that is emitted is $CO_2$ and in the field emission of $N_2O$ is the major contributor

(Rao et al. 2019). Chemical fertilizers in sunflower farming contributed 5.77% of the total carbon footprint (Yousefi et al. 2017). Nitrogenous fertilizer contributed 14% in rice production, while higher contribution is reported under wheat (28%) and maize (39%) in China (Zhang et al. 2017). Again, in rice–mustard cropping system, conventional tillage with residual incorporation and no-tillage with residue retention emitted 33 and 37%, respectively, under fertilizer application (Yadav et al. 2018). Gan et al. (2011) reported about 57–65% share of production and use of nitrogenous fertilizer in total emission in Canadian prairies. Devakumar et al. (2018) also reported that the contribution of inorganic nitrogenous fertilizers is approximately 72 and 9% from phosphorus and potassium fertilizers in semiarid areas of India. Contribution of Asia is highest in emission from fertilizers, followed by America and Europe during 2000–2010 (Tubiello et al. 2013). Fertilizers use in cotton production led to 442 kg $CO_2$-e t$^{-1}$ lint, accounting approximately 57% from on-farm emission, whereas production of fertilizers led to 267.7 kg $CO_2$-e t$^{-1}$ lint, contributing 16.7% in total carbon footprint in Australia (Hedayati et al. 2019). Yue et al. (2017) observed 37–88% contribution of fertilizers in agriculture footprint of different crops. Farag et al. (2013) reported that the contribution of N fertilizers accounts for 10% in rice cultivation in Egypt. For wheat, maize, and soybean crops, the major contribution in carbon footprint is due to use of fertilizers that is more than 75% (Cheng et al. 2015). N fertilizers contributed 89% of carbon footprint, whereas phosphorus (4%) and potassium (2%) contribution were very small in India during 2000–2010 (Sah and devakumar 2018). During the same time period, pesticides accounted for 2% of the total carbon footprint (Sah and devakumar 2018).

**Crop**: Carbon footprint of different crops varied according to the demand of nutrients and management practices. Emission from crops largely depends on the amount of fertilizer used (Gan et al. 2011). Rice is the most energy demanding crop and thus also contributes most in GHG emissions (Rao et al. 2019). Zhang et al. (2017) reported rice as having the highest carbon footprint of 1.60 kg $CO_2$-e per unit yield due to emission of $CH_4$ that contributes 45% of total carbon footprint. Wheat had a lower carbon footprint than rice but higher than maize. The carbon footprint of maize was estimated as 0.48 kg $CO_2$-e per unit yield and for wheat 0.75 kg $CO_2$-e per unit yield in China (Zhang et al. 2017). FAO (2017) report suggested that rice cultivation produces 523 million tons of $CO_2$-e per year that contributed 8.8–10% in 2012 of total agricultural emission globally. In 2015, rice cultivation emitted 2917 Gg $CO_2$-e and wheat emitted 1537 Gg $CO_2$-e which contributed, respectively, 60 and 31% of total cropland emission, in which rice covered 37.7% and wheat covered 44.5% of total cropland area in Punjab, India (Benbi 2018). Rice cropping in India contributed 21% of total agricultural emission (INCCA, Indian Network for Climate Change Assessment 2010). Devakumar et al. (2018) analyzed different crop groups and concluded that oilseeds and commercial crops have a highest carbon footprint of 30 and 29%, respectively, followed by cereals and pulses that contributed around 25 and 16%, respectively. Thus, leguminous crops have the highest sustainability index (Devakumar et al. 2018). Oilseed cropping emits higher GHGs than cereals as they have high N content (Liu et al. 2016). In contrast, Sah and Devakumar (2018) reported highest carbon footprint of cereals followed by oilseeds and then pulses.

Gan et al. (2011) reported highest carbon footprint of canola, followed by mustard, flaxseed, spring wheat, chickpea, dry pea, and then lentil. While comparing carbon footprint of various crops in three soil types, viz. brown, dark brown, and black soil in semiarid regions, higher carbon footprint of crops was recorded in humid black soil. Carbon footprint of different crop plant studied is given in Table 1. Leguminous crops have 65% less emission than canola and wheat (Gan et al. 2011). Kharif cropping has a higher carbon footprint than rabi as the rabi cropping is confined to the areas where irrigation can be done easily (Devakumar et al. 2018). According to Yue et al. (2017), vegetables, among the grains, oilseeds, fruits, etc., have the lowest carbon footprint of 0.15 kg $CO_2$-e $kg^{-1}$. Yadav et al. (2018) reported that carbon footprint due to contribution of $N_2O$ emitted from fertilizer application, mulching, and roots is highest and contributed 41 and 36% for no-tillage and conventional tillage, respectively, in rice–mustard cropping system.

The studies concluded by Rao et al. (2019), Zhang et al. (2017), Benbi (2018), and Cheng et al. (2015) in different regions of the world showed that the highest carbon footprint is recorded by rice due to large-scale emission of $CH_4$. Rao et al. (2019) compared carbon footprint of rice, wheat, sorghum, maize, pearl millet, and finger millet, in all states of India, and found that rice crop has higher energy demand, especially for irrigation. Although for crops, other than rice and wheat, requirement of energy for irrigation is very low and carbon footprint is mainly due to fertilizers and machinery. Millets have the lowest carbon footprint as the crop water requirement is too less. The part of North India shows higher carbon footprint as rice and wheat are there as major crops. Carbon footprint also depends upon water pump efficiency. The efficiency of water pump in India is quite lower (40–50%), whereas internationally, it is 70–90% efficient. Crop cultivation in India is mainly based on water availability and climate conditions of that region. Carbon footprint of different crops in rainfed areas reduces overall carbon footprint due to low carbon footprint from irrigation.

Pandey et al. (2013) studied the effect of tillage on carbon footprint of rice–wheat system, in India. In the study, the comparison of four treatments, tillage before both crops, tillage before rice only, tillage before wheat only, and no-tillage, was done. Tillage before both rice and wheat cropping increased $CO_2$ and $N_2O$ emission but $CH_4$ emission was significantly declined. Therefore, the overall carbon footprint of this treatment was found to be negative. In no-tillage condition, the yield response was better, but carbon footprint was also high. Tillage before rice transplantation and no-tillage before wheat sowing showed highest carbon footprint among all treatments.

Cheng et al. (2015) analyzed the national statistical data of major crops such as rice, wheat, maize, and soybean for carbon footprint calculation of China in the year 2011. The highest carbon footprint was found for rice (0.37 kg $CO_2$-e $kg^{-1}$), followed by wheat (0.14 kg $CO_2$-e $kg^{-1}$) and maize (0.12 kg $CO_2$-e $kg^{-1}$), and lowest was recorded for soybean (0.10 kg $CO_2$-e $kg^{-1}$). Reduced carbon footprint of soybean may be attributed to the reduced use of N fertilizer. Carbon footprint of these crops was positively correlated with N fertilization rate in production with $r^2$ value of about 0.91, whereas for rice cropping, carbon footprint was correlated with $CH_4$ emission with $r^2$ value of 0.85 (Cheng et al. 2015). Variation in carbon footprint for the same

**Table 1** Carbon footprint of selected crops

| S. No. | Crops | Carbon footprint (kg $CO_2$-e $kg^{-1}$) | Comments | References |
|---|---|---|---|---|
| 1 | Canola | 0.913 | Pre- and on-farm CF (carbon footprint) of crops grown in brown soil of Canadian prairie | Gan et al. (2011) |
| | Mustard | 0.601 | | |
| | Flaxseed | 0.456 | | |
| | Chickpea | 0.254 | | |
| | Dry pea | 0.189 | | |
| | Lentil | 0.164 | | |
| | Spring wheat | 0.383 | | |
| 2 | Cotton | 1.600 | Pre-, on-, and post-farm CF in Australia | Hidayati et al. (2019) |
| 3 | Rice | 0.37 | Pre- and on-farm CF in China | Cheng et al. (2015) |
| | Wheat | 0.14 | | |
| | Maize | 0.12 | | |
| | Soybean | 0.10 | | |
| 4 | Rice | 1.90 | On-farm CF in Egypt | Farag et al. (2013) |
| 5 | Maize | 0.21–0.24 | On-farm CF in India | Jat et al. (2019) |
| 6 | Tomato | 0.11 | On-farm CF in China | Yue et al. (2017) |
| | Cucumber | 0.14 | | |
| | Eggplant | 0.18 | | |
| | Green pepper | 0.33 | | |
| 7 | Rice | 1.60 | Pre- and on-farm CF in China | Zhang et al. (2017) |
| | Wheat | 0.75 | | |
| | Maize | 0.48 | | |
| 8 | Sunflower | 0.875 | On-farm CF in Iran | Yousefi et al. (2017) |
| 9 | Rice | 0.80 | On-farm CF in China | Yan et al. (2015) |
| | Wheat | 0.66 | | |
| | Maize | 0.33 | | |
| 10 | Wheat | 1.061 | On-farm CF in India in no-tillage condition | Pandey et al. (2013) |

crop was also recorded for different regions in China due to different N application rate and irrigation.

Yousefi et al. (2017) estimated the carbon footprint of sunflower agroecosystem in Iran. Data collection was done from 70 sunflower agroecosystems in 2012, and mean carbon footprint was found to be 0.875 kg $CO_2$-e $kg^{-1}$. The output energy of sunflower was higher than the input energy. Major share in input energy was of electricity that was used to power water pump. Sunflower grows in Iran during summer, so water loss due to evapotranspiration is high. Sah and Devakumar (2018) also reported sunflower as having higher carbon footprint.

Zhang et al. (2017) assessed carbon footprints of rice, wheat, and maize in different regions of China through a survey that started in 2014. The study showed similar results as of Rao et al. (2018) and Cheng et al. (2015). Rice was found to have the highest carbon footprint, i.e., 2.10 kg $CO_2$-e $kg^{-1}$, followed by wheat (CF-1.26 kg $CO_2$-e $kg^{-1}$) and then maize (CF-0.95 kg $CO_2$-e $kg^{-1}$). The farm emission, manufacturing, processing and transport of fertilizer, electricity, chemicals, and machinery were also included in total carbon footprint. The study showed a higher contribution of later processes rather than the on-farm emission. For wheat cropping, the share of upstream input is 86% and is highest; among the studied crops, followed by maize, the contribution is 63% and then rice, for which the contribution of upstream inputs is 50%. Differences in carbon footprint of the same crop in different regions are due to regional differences, nutrient requirements, irrigation, and farming practices.

Straw burning increases the carbon footprint of crops significantly. Powlson et al. (2016) reported a great loss of C due to burning of straw in field and also as the fuel. The carbon stored in residues is lost in atmosphere totally as $CO_2$. Farag et al. (2013) estimated 35.8% share of rice straw burning in total crop production emission in Egypt during 2008–2011. Burning of crop residues not only emits GHGs but also emits many harmful gases that create negative effects on human as well as the environment. Sahai et al. (2011) reported 80–85% of GHGs due to field burning of rice, wheat, and sugarcane residues. In India, 488 million tons of crop residues was generated in 2017 and 24% of it get burnt, which emitted 211 Tg $CO_2$-e GHGs, along with other gaseous air pollutants (Ravindra et al. 2019). The major contribution is of $CO_2$, followed by $CH_4$ and $N_2O$ (Ravindra et al. 2019).

**Livestock**: Livestock sector emits GHGs majorly via enteric fermentation, feed production, transport, and manure application. Carbon footprint from livestock is more than doubled compared to the crops. Globally, $CH_4$ is the dominant GHG emitted from livestock sector, which contributed about 44% of total, whereas $N_2O$ and $CO_2$ account for approximately similar contribution, i.e., 29 and 27%, respectively (Gerber et al. 2013). Globally, livestock contributed about 66% of total agricultural GHG emissions, in which enteric fermentation accounts for 40%, manure left in pastures accounts for 16%, and manure management share is 7% (FAOSTAT 2014) (Fig. 2). Gerber et al. (2013) reported that livestock contributed 14.5% of total anthropogenic emission globally. From India, Chhabra et al. (2013) reported 99.8% $CH_4$ and 0.2% $N_2O$ emission from livestock. From livestock sector, feed production and transport contributed 45% and the second largest share is by enteric fermentation which is approximately 40% of the total livestock sector (Gerber et al. 2013).

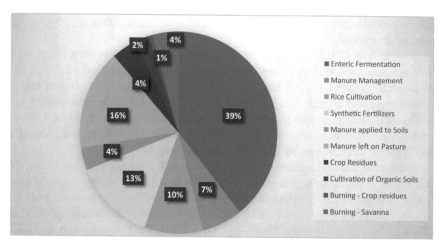

**Fig. 2** Global emission of GHG from agriculture sector (FAOSTAT 2014)

Also, the contribution of beef and dairy cattle is more in carbon footprint than other ruminants or animals (Gerber et al. 2013; Chhabra et al. 2013). Cattles sharing the major contributions in GHG emissions are from the production of milk and meat. From pork and poultry, the emission is mainly from fed production, processing, and manure processing. The variations in carbon footprint of different livestock are also due to variations in management practices, environmental conditions, as well as on processing and marketing. By manure management, major share is of $N_2O$ and $CH_4$, whereas $CO_2$ and $N_2O$ are emitted from feed processing and $CO_2$ and $CH_4$ from energy consumption and enteric fermentation, respectively. Enteric fermentation and manure management in India account for 91 and 9%, respectively, of the total $CH_4$ emission (Chhabra et al. 2013). Manure contains organic matter and N that lead to $CH_4$ emission by decomposition of organic matter and $N_2O$ emission indirectly by $NH_3$ emission. Chhabra et al. (2013) reported that the livestock footprint of India in 2003 was 247.2 Mt $CO_2$-e and $CH_4$ emission from livestock was 74.4 kg ha$^{-1}$. INCCA (2010) report suggests that enteric fermentation is responsible for 63% of total emission from agriculture sector, whereas manure management contributed 0.73%, during 2007. Gerber et al. (2013) reported lower emission from buffalo, whereas higher emission from buffalo due to high demand of meat and milk in India was reported by Chhabra et al. (2013). Yue et al. (2017) reported livestock-based products have higher carbon footprint than crop-based products. The carbon footprints of meat, poultry egg, and milk were reported as 6.21, 4.09, and 1.47 kg $CO_2$-e kg$^{-1}$, respectively, which were higher than the crops, vegetables, oilseeds, soybeans, and fruits, having the carbon footprint of less than 1 kg $CO_2$-e kg$^{-1}$. The higher carbon footprint of animal-based products is due to higher carbon footprint of animal feed.

# 4   Mitigation Strategies

Under the current scenario of adverse effects of climate change, it is very important to reduce the carbon footprint of agricultural products. Awareness of carbon footprint of the products is important in local people, so the choice of preferences would help to reduce the emission. Figure 3 shows the practices that can be utilized to reduce GHG emissions.

**Crop diversification**: Carbon footprint of agricultural products varies with species to species and various agricultural practices. Diversified cropping system increases the productivity as well as lowers the carbon footprint (Gan et al. 2011; Liu et al. 2016). Leguminous crops have lower carbon footprint by sequestering carbon and nitrogen (Gan et al. 2011; Liu et al. 2016). Nitrogenous fertilizers during crop cultivation are one of the major sources of agricultural emission of GHGs. The crops grown after leguminous crops require less N fertilization, so total carbon footprint is lowered. Even the requirement of energy for manufacture and transportation of fertilizers is reduced, leading to low GHG emissions. Hedayati et al. (2019) reported reduction in total GHG emission by 13.2% by using optimum rate of N fertilizers. Lentil wheat rotation has the same yield as of continuous wheat cropping but with low N addition and thus improves the N use efficiency. Lentil and wheat system was reported to have lowest carbon footprint compared to continuous wheat, fallow flax wheat, and fallow wheat systems as carbon footprint was reduced by 150% compared to wheat monoculture (Gan et al. 2014).Cropping sequence in a crop rotation system also accounted immensely in the carbon footprint (Gan et al. 2011; Liu et al. 2016). Gan et al. (2011) reported wheat crop when planted after leguminous crops emitted 20% lower GHGs compared to when cultivated after cereals. Similarly, when grown after oilseed, the emission of GHGs from wheat cropping reduced by 11% as compared to cereals. Crop biomass and N content of the crop are also the responsible factors that decide the carbon footprint.

**Summer fallowing**: Summer fallowing system reduces the carbon footprint of agriculture as it increases the N availability and thus reduces the amount of N fertilizer

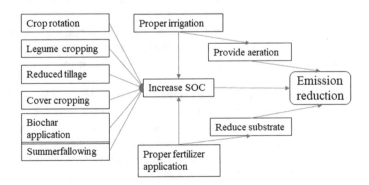

**Fig. 3**  Mitigation measures to reduce carbon footprint from agriculture

required. Summer fallowing also enhances the productivity (Liu et al. 2016). In contrast, Gan et al. (2014) reported reduced frequency of summer fallowing for reducing the carbon footprint. Instead of summer fallowing, cropping of legumes reduced the carbon footprint to more extent. Residue removal although reported to emit higher $CO_2$ and also causes the loss of SOC, it reduces the emission of $N_2O$ (Liska et al. 2014).

**Increasing SOC**: Soil acts as a sink for organic matter and practices that directly releases terrestrial carbon to atmosphere, such as burning of straw and fossil fuel, leading to loss of SOC and increase the GHG concentration. For increasing the SOC, conservation tillage, integrated nutrient management, mulching, cover cropping, diverse cropping system, and biochar application are recommended (Lal 2011). Deficit irrigation helps to increase SOC and thus can be used to reduce carbon footprint as the allocation of biomass is more toward below ground in order to access more water (Pawlowski et al. 2017). Pressurized irrigation systems can help in lessen the total carbon footprint from agriculture (Hedayati et al. 2019). Powlson et al. (2016) concluded that conservation agriculture enhances the SOC through reduced tillage, incorporation of crop residues, and diversification of crops through a meta-analysis in Indo-Gangetic plain and Sub-Saharan Africa.

**Mitigation during rice cultivation**: Weller et al. (2015) reported that flooded rice emits 90% higher $CH_4$ than the aerobic rice, but $N_2O$ emission from flooded rice system was lower insignificantly. Emission of $CH_4$ from rice fields can be lowered by rotation of rice with aerobic rice or maize in dry season and with rice in wet season although $N_2O$ emission was reported to be enhanced (Janz et al. 2019). Proper anaerobic condition is not maintained in aerobic rice as flooding of rice is not done as well as drying of soil occurs in between irrigations. So, during aeration of soil, methanogens are unable to produce $CH_4$ and methanotrophs are encouraged for $CH_4$ oxidation (Janz et al. 2019). Rotation with maize reduced the emission as maize cropping acted as weak sink for $CH_4$ (Janz et al. 2019; Linquist et al. 2012). From rice–aerobic rice and rice–maize rotation, although $N_2O$ emission was higher from straw application, GWP of these rotations was 33% and 48%, respectively, lower than rice–rice rotation (Janz et al. 2019). Rice maize rotation also reduces the irrigation water requirement, so overall carbon footprint is greatly reduced (Janz et al. 2019). Yao et al. (2017) suggested the production of rice in ground cover rice production system, where rice crop is covered with a thin plastic sheet, so that the moisture level is maintained and thus direct emission from flooding as well as carbon footprint of irrigation will be reduced. This method reduces the total carbon footprint along with increasing the yield.

**Biochar application**: Biochar helps to sequester carbon in soil (Gan et al. 2011; Lal 2011; IPCC 2014; Zhang et al. 2012). Zhang et al. (2012) found a reduction in emission after biochar addition, in China. However, different concentrations of biochar did not have any significant effect on GHG emissions, applied different concentrations of biochar to the soil in rice field in China. Biochar application reduced the GHG emission and also enhanced the crop yield (Xiao et al. 2019). Sun et al. (2019) also reported reduced emission after biochar application in rice field. Biochar application in soil causes alteration of soil biota and improves soil characteristics that lead to lower emission of $CH_4$ and $N_2O$. Sun et al. (2019) speculated that applied biochar enhanced the methanotrophic community or decreased the population of methanogens, enhanced N immobilization, decreased denitrification, and increased soil pH, leading to reduced $CH_4$ and $CO_2$ emission. Xiao et al. (2019) suggested reduction in $CH_4$ and $N_2O$ emission by applying biochar in rice field under controlled irrigation.

**Organic farming**: In organic farming system, energy use is greatly reduced due to no use of fertilizers and pesticides. Carbon sequestration is high in organic farming. Organic farming helps in reducing the total carbon footprint of agriculture (Meier et al. 2015; Adewale et al. 2018; Skinner et al. 2019; Hedayati et al. 2019). Manure composting done in organic farming can reduce $N_2O$ emission by 50% and $CH_4$ emission by 70% (IFOAM 2016). Skinner et al. (2019) reported that organic farming reduces the $N_2O$ emission by 40.2%. Proper management of manure also helps in lessen the carbon footprint. IFOAM (2016) report suggested that turning and aeration of manure heaps, keeping manure in closed space, solid and slurry separation and addition of substances that reduce the emission and biogas production by biodigesters from manure, can be used to lower the emission from manure management. N content in animal feed in the form of crude protein is also responsible for emission. Concentrates in diet lead to higher $CH_4$ emission (Meier et al. 2015). This is because it increases the undigested part in the manure and degradation of the undigested matter emits $CH_4$.

**Biofuel**: Use of biofuels in place of fossil fuels is also reported to reduce the carbon footprint of agriculture (Pawlowski et al. 2017). Crop residues containing lignin can be burnt to produce biofuel that reduces the overall emission from electricity used (Liska et al. 2014). Solar powered irrigation pumps reduced the total agricultural carbon footprint by 8.1% by substituting the electrical energy and also the use of biofuel-based machinery instead of diesel-based reduced the emission by 3.9% in cotton cultivation (Hedayati et al. 2019). Unblended biodiesel can also be used that is generated from wastes. Pawlowski et al. (2017) reported sugarcane and napier grass grown in place of arable crops, reduced the carbon footprint, when grown with conservation farming practices. It also helped in increasing SOC. Napier grass cultivation has more environmental benefits in terms of reduced GHG emissions, increase in SOC and biofuel production.

**Tillage**: Tillage causes the disturbances in soil thus carbon stabilization in soil gets disturbed, organic matters are unveiled for the oxidation and thus loss of SOC occurs. Mulching of crop residue with no-tillage significantly enhanced the soil carbon and also stabilizes the new aggregates. Reduced tillage causes increment in total and recalcitrant C pool in rice–wheat system (Pandey et al. 2014). Crop residues left under no-tillage condition add organic carbon to the soil (Pandey et al. 2014; Powlson et al. 2016; Nath et al. 2017; Yadav et al. 2018), and also, the rate of oxidation of organic molecules is greatly reduced due to soil cover (Lal 2004). Carbon sequestration rate was found highest under continuous no-tillage condition by Pandey et al. (2014). No-tillage reduces the $CH_4$ and $N_2O$ emission but enhances $CO_2$ emission, but the GWP is reduced (Pandey et al. 2012). Similarly, Nath et al. (2017) found that during no-tillage condition, soil moisture content and soil temperature are higher, thus making the conditions for denitrification more favorable. In contrast, Powlson et al. (2016) observed that no-tillage may increase the C sequestration, when the decomposition rate is slow and input C is also stabilized. Pandey et al. (2013) also reported lowest carbon footprint under tillage in rice–wheat cropping system. The potential of no-tillage to enhance the SOC depends upon the region and soil condition.

**N fertilizer**: By the proper application of N-based fertilizer, mitigation of emission can be done (Yue et al. 2017). By the limited use of fertilizers, emission from soil as well as emission from production and transportation may also be reduced. Even a small change in applied N fertilizers can lead to a big difference in emission pattern. Various techniques are developed to apply only the required amount of fertilizers such as GreenSeeker and leaf color chart-based urea application. Nath et al. (2017) reported lower emission of $N_2O$ by the use of GreenSeeker by 11–13% in wheat cropping. International Rice Research Institute developed a more efficient technique to reduce the excessive use of N fertilizer that is based on the color of leaf. Urea is applied by comparing a chart provided, to the color of leaf, and is called leaf color chart (LCC)-based urea application (Bhatia et al. 2012). This LCC-based urea application method can be used to reduce the fertilizer-based emission, and it also enhances the N use efficiency as well as yield of the crop (Bhatia et al. 2012). Bhatia et al. (2012) reported 10.5% reduction in GWP of rice–wheat system through using LCC-based urea application. Jat et al. (2019) suggested the use of neem-coated urea. Jat et al. (2019) compared neem-coated and sulfur-coated urea from prilled urea in maize cropping system in conservation agriculture and concluded that coated urea not only lowered the carbon footprint of maize cropping, but also enhanced the yield. Although carbon footprint was recorded lowest when no N fertilizer was used, but significant yield loss occurred. Deep placement of urea rather than surface application is reported to reduce the GHG emission by 41% under ground cover rice production system (Yao et al. 2017). For the deep placement of urea, it is placed in 10–15-cm deep holes made near each rice hills. To reduce the carbon footprint, different methods can be applied to fields that will prevent excess use of fertilizers and also improve the soil quality.

## 5   Models to Estimate Carbon Footprint

To mitigate the emission from agriculture in future, various models are developed and used. As the climate change is very diverse and carbon footprint of agriculture sector is complex, so the developed models help to assess the impact and mitigation strategies for agricultural system. Emission factor methods are also used for carbon footprint estimation of variety of crops such as wheat, canola, maize, and sunflower. Emission factor is simpler models, and it uses IPCC tier I methodology. Limitations with emission factor methodology are: (1) It is region specific; (2) less interaction of soil, climate, and crop management. Again, some models are simple that use only biomass or yield or soil carbon or other properties of soil or manure input, etc.,

**Table 2** Models for assessing soil C dynamics influencing carbon footprints and their required inputs

| Models | Inputs |
|---|---|
| Roth C | Soil temperature, soil water, clay content |
| C-TOOL | Mean monthly air temperature, clay content, C/N ratio, C in organic inputs |
| ICBM | Crop type, soil temperature, rainfall, soil characteristic, tillage frequency |
| DayCent | Daily min/max temperature and precipitation, soil texture, vegetation type, amount and timing of nutrient amendment |
| DNDC | Site and climate, crop, tillage, fertilizer and manure amendment, plastic film use, flooding, irrigation, grazing, and cutting |
| CERES-EGC | Weather, soil properties, crop management, soil organic matter |
| Info-RCT | Precipitation, manure/residue application, SOC, human labor, animal labor, machine, seed |

Abbreviations *Roth C* Rothamsted Carbon, *ICBM* Introductory Carbon balanced model, *DayCent* the daily time step version of CENTURY, *DNDC* DeNitrification DeComposition, *CERES-EGC* Crop Environment REsource Synthesis Environnement et Grandes cultures, *Info-RCT* Information on Use of Resource Conservation Technologies

and some complex models are dynamic crop–climate–soil models. Roth C (Rothamsted Carbon, Coleman and Jenkinson 1996), ICBM (Introductory Carbon Balanced Model, Andrèn and Kätterer 1997) and C-TOOL (Hamelin et al. 2012) are few of the simple C models. In these models, soil characteristics are included as input such as soil temperature, water content and clay content, and crop type (Table 2). These models give soil C change, along a time series. Roth C uses monthly time step, and C-TOOL uses both daily and annual time steps. These models are simple and easy to use. Although these models have some drawbacks as they are not applicable globally, they did not take into account other determinants and all the soil borne emissions are not included.

The more complex and mathematical models are dynamic crop–climate–soil models. The output of these models is shown as soil C and N, crop growth, and GHG emission. CERES-EGC (Crop Environment REsource Synthesis Environnement et Grandes cultures, Gabrielle and Gagnaire 2008), DNDC (DeNitrification DeComposition), CropSyst (Cropping Systems Simulation Model, Zaher et al. 2013), and DayCent (the daily time step version of CENTURY, Kim and Dale 2009) models estimate change in SOC stock as well as GHG emission. These models can be used for agricultural fields, grassland, and forests and require large statistical data. Saharawat et al. (2012) used Info-RCT (Information on Use of Resource Conservation Technologies) model that predicted crop yield, N budget, and GHG emission in South Asian region, developed by Pathak et al. (2011). The models to be used are selected depending upon the objective and data availability. Goglio et al. (2015) suggested that for small-scale site-specific management, dynamic crop–climate–soil model should be preferred then simple models. Similar is with large-scale assessment, so dynamic crop–climate–soil model is better and can be used in all cases, but it is cumbersome.

## 6  Conclusion

Agriculture shares a major proportion in impacting the climate change scenario via a higher carbon footprint. Processes related to agriculture from the production of agricultural input to the processing of agricultural output and emission of GHGs are inevitable. From the studies, it can be concluded that carbon footprint of pre-farm activities such as manufacture and transportation of fertilizers, pesticides, and the machineries is significant. Emission from field is largely dependent on crop type and their water and fertilizer requirements. Rice crops have the highest carbon footprint due to $CH_4$ emission and irrigation demand, among all the studied crops. Carbon footprint of livestock sector is quite higher due to manure management, feed production, and enteric fermentation. The interrelation of the processes decides the carbon footprint, and thus, mitigation of emission can be done by using some improved agricultural practices. For mitigation, appropriate use of fertilizer, crop rotation, irrigation management, biochar application, reduced tillage frequency, organic farming, etc., are some suggested measures. Farming practices that enhance SOC are the best mitigation strategies for lowering the carbon footprint. Carbon footprint of N fertilizer is observed highest among all, so various techniques are developed such as GreenSeeker, LCC-based urea application, neem, and sulfur-coated urea to reduce the carbon footprint of N fertilizers. Various models are developed to estimate carbon footprint that take some physicochemical properties of soil, crop data, and agricultural input data to provide SOC dynamics and GHG emission. Models should not be used only for carbon footprint calculation, but also as a tool to foresee the positive effects of management practices.

**Acknowledgements** Authors are thankful to APN project (CRRP2016-09MY-Lokupitiya) for financial support. Bhavna Jaiswal is thankful to University Grant Commission (UGC), New Delhi, for Junior Research Fellowship.

## References

Adewale C, Reganold JP, Higgins S, Evans RD, Carpenter-Boggs L (2018) Improving carbon footprinting of agricultural systems: Boundaries, tiers, and organic farming. Environ Impact Assess Rev 71:41–48

Andren O, Kätterer T (1997) ICBM: the introductory carbon balance model for exploration of soil carbon balances. Ecol Appl 7:1226–1236

Benbi DK (2018) Carbon footprint and agricultural sustainability nexus in an intensively cultivated region of Indo-Gangetic Plains. Sci Total Environ 644:611–623

Bhatia A, Pathak H, Jain N, Singh PK, Tomer R (2012) Greenhouse gas mitigation in rice–wheat system with leaf color chart-based urea application. Environ Monit Assess 184(5):3095–3107

Cheng K, Yan M, Nayak D, Pan GX, Smith P, Zheng JF, Zheng JW (2015) Carbon footprint of crop production in China: an analysis of National Statistics data. J Agric Sci 153(3):422–431

Chhabra A, Manjunath KR, Panigrahy S, Parihar JS (2013) Greenhouse gas emissions from Indian livestock. Clim Change 117(1–2):329–344

Coleman K, Jenkinson DS (1996) RothC-26.3-A model for the turnover of carbon in soil. In: Evaluation of soil organic matter models. Springer, Berlin, Heidelberg, pp 237–246

Devakumar AS, Pardis R, Manjunath V (2018) Carbon footprint of crop cultivation process under semiarid conditions. Agric Res 7(2):167–175

FAO (2014) Food and agriculture organization of the United Nations: News article assessed on http://www.fao.org/news/story/en/item/216137/icode/

FAO (2015) Food and agriculture organization of the United Nations: FAO statistical pocketbook, 2015. http://www.fao.org/3/a-i4691e.pdf

FAO (2017) FAOSTAT database collections. Food and Agriculture Organization of the United Nations. Rome. http://faostat.fao.org

FAOSTAT (2014). Food and agriculture organization of the United Nations http://www.fao.org/3/a-i3671e.pdf

FAOSTAT (2019) Food and agriculture organization of the United Nations http://www.fao.org/faostat/en/#home

Farag AA, Radwan HA, Abdrabbo MAA, Heggi MAM, McCarl BA (2013) Carbon footprint for paddy rice production in Egypt. Nat Sci 11(12):36–45

Gabrielle B, Gagnaire N (2008) Life-cycle assessment of straw use in bio-ethanol production: a case study based on biophysical modelling. Biomass Bioenerg 32(5):431–441

Gan Y, Liang C, Chai Q, Lemke RL, Campbell CA, Zentner RP (2014) Improving farming practices reduces the carbon footprint of spring wheat production. Nat Commun 5:5012

Gan Y, Liang C, Hamel C, Cutforth H, Wang H (2011) Strategies for reducing the carbon footprint of field crops for semiarid areas. A Rev Agron Sustain Dev 31(4):643–656

Gerber PJ, Steinfeld H, Henderson B, Mottet A, Opio C, Dijkman J, Falcucci A, Tempio G (2013) Tackling climate change through livestock: a global assessment of emissions and mitigation opportunities. Food and Agriculture Organization of the United Nations (FAO)

Goglio P, Grant BB, Smith WN, Desjardins RL, Worth DE, Zentner R, Malhi SS (2014) Impact of management strategies on the global warming potential at the cropping system level. Sci Total Environ 490:921–933

Goglio P, Smith WN, Grant BB, Desjardins RL, McConkey BG, Campbell CA, Nemecek T (2015) Accounting for soil carbon changes in agricultural life cycle assessment (LCA): a review. J Clean Prod 104:23–39

Hamelin L, Jørgensen U, Petersen BM, Olesen JE, Wenzel H (2012) Modelling the carbon and nitrogen balances of direct land use changes from energy crops in Denmark: a consequential life cycle inventory. Gcb Bioenergy 4(6):889–907

Harris NL, Brown S, Hagen SC, Saatchi SS, Petrova S, Salas W et al (2012) Baseline map of carbon emissions from deforestation in tropical regions. Science 336(6088):1573–1576

Hedayati M, Brock PM, Nachimuthu G, Schwenke G (2019) Farm-level strategies to reduce the life cycle greenhouse gas emissions of cotton production: An Australian perspective. J Clean Prod 212:974–985

IFOAM report 2016 https://www.ifoameu.org/sites/default/files/ifoameu_advocacy_climate_change_report_2016.pdf

INCCA (Indian Network for Climate Change Assessment), (2010, May). India: greenhouse gas emissions 2007. Ministry of environment and forests, government of India

IPCC (2014) Fifth assessment report of the intergovernmental panel on climate change http://www.ipcc.ch/report/ar5/

Janz B, Weller S, Kraus D, Racela HS, Wassmann R, Butterbach-Bahl K, Kiese R (2019) Greenhouse gas footprint of diversifying rice cropping systems: Impacts of water regime and organic amendments. Agr Ecosyst Environ 270:41–54

Jat SL, Parihar CM, Singh AK, Kumar B, Choudhary M, Nayak HS, Parihar MD, Parihar N, Meena BR (2019) Energy auditing and carbon footprint under long-term conservation agriculture-based intensive maize systems with diverse inorganic nitrogen management options. Sci Total Environ 664:659–668

Kim S, Dale BE (2009) Regional variations in greenhouse gas emissions of biobased products in the United States—corn-based ethanol and soybean oil. Int J Life Cycle Assess 14(6):540–546

Lal R (2004) Carbon emission from farm operations. Environ Int 30(7):981–990

Lal R (2011) Sequestering carbon in soils of agro-ecosystems. Food Policy 36:S33–S39

Linquist B, Van Groenigen KJ, Adviento-Borbe MA, Pittelkow C, Van Kessel C (2012) An agronomic assessment of greenhouse gas emissions from major cereal crops. Glob Change Biol 18(1):194–209

Liska AJ, Yang H, Milner M, Goddard S, Blanco-Canqui H, Pelton MP, Fang SS, Zhu H, Suyker AE (2014) Biofuels from crop residue can reduce soil carbon and increase $CO_2$ emissions. Nat Clim Chang 4(5):398

Liu C, Cutforth H, Chai Q, Gan Y (2016) Farming tactics to reduce the carbon footprint of crop cultivation in semiarid areas. A Rev Agron Sustain Dev 36(4):69

Meier MS, Stoessel F, Jungbluth N, Juraske R, Schader C, Stolze M (2015) Environmental impacts of organic and conventional agricultural products–Are the differences captured by life cycle assessment? J Environ Manage 149:193–208

Nath CP, Das TK, Rana KS, Bhattacharyya R, Pathak H, Paul S, Meena MC, Singh SB (2017) Greenhouse gases emission, soil organic carbon and wheat yield as affected by tillage systems and nitrogen management practices. Arch Agron Soil Sci 63(12):1644–1660

Pandey D, Agrawal M (2014) Carbon footprint estimation in the agriculture sector. In: Assessment of Carbon Footprint in Different Industrial Sectors, vol 1. Springer, Singapore, pp 25–47

Pandey D, Agrawal M, Bohra JS (2012) Greenhouse gas emissions from rice crop with different tillage permutations in rice–wheat system. Agric, Ecosyst Environ 159:133–144

Pandey D, Agrawal M, Bohra JS (2013) Impact of four tillage permutations in rice–wheat system on GHG performance of wheat cultivation through carbon footprinting. Ecol Eng 60:261–270

Pandey D, Agrawal M, Bohra JS, Adhya TK, Bhattacharyya P (2014) Recalcitrant and labile carbon pools in a sub-humid tropical soil under different tillage combinations: a case study of rice–wheat system. Soil Tillage Res 143:116–122

Pathak H, Saharawat YS, Gathala M, Ladha JK (2011) Impact of resource-conserving technologies on productivity and greenhouse gas emissions in the rice-wheat system. Greenh Gases: Sci Technol 1(3):261–277

Patthanaissaranukool W, Polprasert C (2016) Reducing carbon emissions from soybean cultivation to oil production in Thailand. J Clean Prod 131:170–178

Pawlowski MN, Crow SE, Meki MN, Kiniry JR, Taylor AD, Ogoshi R, Youkhana A, Nakahata M (2017) Field-based estimates of global warming potential in bioenergy systems of hawaii: crop choice and deficit irrigation. PLoS ONE 12(1):e0168510

Powlson DS, Stirling CM, Thierfelder C, White RP, Jat ML (2016) Does conservation agriculture deliver climate change mitigation through soil carbon sequestration in tropical agro-ecosystems? Agric Ecosyst Environ 220:164–174

Rao ND, Poblete-Cazenave M, Bhalerao R, Davis KF, Parkinson S (2019) Spatial analysis of energy use and GHG emissions from cereal production in India. Sci Total Environ 654:841–849

Ravindra K, Singh T, Mor S (2019) Emissions of air pollutants from primary crop residue burning in India and their mitigation strategies for cleaner emissions. J Clean Prod 208:261–273

Rees WE (1992) Ecological footprints and appropriated carrying capacity: what urban economics leaves out. Environ Urban 4:121–130

Sah D, Devakumar AS (2018) The carbon footprint of agricultural crop cultivation in India. Carbon Manag 9(3):213–225

Sahai S, Sharma C, Singh SK, Gupta PK (2011) Assessment of trace gases, carbon and nitrogen emissions from field burning of agricultural residues in India. Nutr Cycl Agroecosyst 89(2):143–157

Saharawat YS, Ladha JK, Pathak H, Gathala MK, Chaudhary N, Jat ML (2012) Simulation of resource-conserving technologies on productivity, income and greenhouse gas GHG emission in rice-wheat system. J Soil Sci Environ Manag 3(1):9–22

Skinner C, Gattinger A, Krauss M, Krause HM, Mayer J, van der Heijden MG, Mäder P (2019) The impact of long-term organic farming on soil-derived greenhouse gas emissions. Sci Rep 9(1):1702

Sun H, Lu H, Feng Y (2019) Greenhouse gas emissions vary in response to different biochar amendments: an assessment based on two consecutive rice growth cycles. Environ Sci Pollut Res 26(1):749–758

Tubiello FN, Salvatore M, Rossi S, Ferrara A, Fitton N, Smith P (2013) The FAOSTAT database of greenhouse gas emissions from agriculture. Environ Res Lett 8(1):015009

Weller S, Kraus D, Ayag KRP, Wassmann R, Alberto MCR, Butterbach-Bahl K, Kiese R (2015) Methane and nitrous oxide emissions from rice and maize production in diversified rice cropping systems. Nutr Cycl Agroecosyst 101(1):37–53

Wiedmann T, Minx J (2008) A definition of 'carbon footprint'. Ecol Econ Res Trends 1:1–11

Xiao YN, Sun X, Ding J, Jiang Z, Xu J (2019) Biochar improved rice yield and mitigated $CH_4$ and $N_2O$ emissions from paddy field under controlled irrigation in the Taihu Lake Region of China. Atmos Environ 200:69–77

Yadav GS, Das A, Lal R, Babu S, Meena RS, Saha P, Singh R, Datta M (2018) Energy budget and carbon footprint in a no-till and mulch-based rice–mustard cropping system. J Clean Prod 191:144–157

Yan M, Cheng K, Luo T, Yan Y, Pan G, Rees RM (2015) Carbon footprint of grain crop production in China-based on farm survey data. J Cleaner Prod 104:130–138

Yao Z, Zheng X, Zhang Y, Liu C, Wang R, Lin S, Zuo Q, Butterbach-Bahl K (2017) Urea deep placement reduces yield-scaled greenhouse gas ($CH_4$ and $N_2O$) and NO emissions from a ground cover rice production system. Sci Rep 7(1):11415

Yousefi M, Khoramivafa M, Damghani AM (2017) Water footprint and carbon footprint of the energy consumption in sunflower agroecosystems. Environ Sci Pollut Res 24(24):19827–19834

Yue Q, Xu X, Hillier J, Cheng K, Pan G (2017) Mitigating greenhouse gas emissions in agriculture: From farm production to food consumption. J Clean Prod 149:1011–1019

Zaher U, Stöckle C, Painter K, Higgins S (2013) Life cycle assessment of the potential carbon credit from no-and reduced-tillage winter wheat-based cropping systems in Eastern Washington State. Agric Syst 122:73–78

Zhang A, Bian R, Pan G, Cui L, Hussain Q, Li L, Zheng J, Zheng J, Zhang X, Han X, Yu X (2012) Effects of biochar amendment on soil quality, crop yield and greenhouse gas emission in a Chinese rice paddy: a field study of 2 consecutive rice growing cycles. Field Crop Res 127:153–160

Zhang D, Shen J, Zhang F, Li YE, Zhang W (2017) Carbon footprint of grain production in China. Sci Rep 7(1):4126

Printed in the United States
By Bookmasters